Frontiers in
Neurobiology 2

Series Editors:
Anthony J. Turner (Leeds)
F. Anne Stephenson (London)

FRONTIERS IN NEUROBIOLOGY 2

Nerve Growth and Guidance

Edited by
C.D. McCaig

PORTLAND PRESS

#35017275

Published by Portland Press Ltd, 59 Portland Place, London W I N 3AJ, U.K.

In North America orders should be sent to Ashgate Publishing Co., Old Post Road, Brookfield, VT 05036-9704, U.S.A.

© 1996 Portland Press Ltd, London

ISBN I 85578 085 2   ISSN 1353-6508

British Library Cataloguing in Publication Data
A catalogue record for this book is available from the British Library

Although, at the time of going to press, the information contained in this publication is believed to be correct, neither the authors nor the editors nor the publisher assume any responsibility for any errors or omissions herein contained. Opinions expressed in this book are those of the authors and are not necessarily held by the editors or the publishers.

Printed in Great Britain by Henry Ling Ltd, Dorchester, U.K.

Contents

Preface

The topic of this second volume in the *Frontiers in Neurobiology* series can truly be said to be a growth area. What has for long been a collection of phenomenological events is now being dissected at the molecular level and the components involved in the growth and guidance of axons are being characterized in detail. Regulation of these neuronal processes, especially through the actions of phosphorylation and second messenger systems, are also beginning to be understood. A particular focus of this volume is the contribution of endogenous electrical fields to the whole process of neuronal growth.

Central to the events mediating axonal growth is the growth cone itself, and several chapters in this volume deal with distinct aspects of the organization, function and regulation of this remarkable structure. Thus Gordon-Weeks and colleagues describe the nature of the growth cone cytoskeleton and the proteins involved in 'pathfinding', including tau and the microtubule-associated proteins which have been implicated in developmental events and 'plasticity'. Among the cell-surface proteins involved in neurite outgrowth are the fascinating family of immunoglobulin-like cell-adhesion molecules; Baldwin et al. summarize critically the evidence that protein tyrosine kinases regulate growth via the neural cell-adhesion molecules. A subsequent chapter by Davenport evaluates the role of intracellular signalling, especially calcium fluxes in dynamic events within the growth cone. This issue is developed further by Bandtlow, who examines the role of neurite inhibitors and calcium in the opposing event, the collapse of the growth cone.

Another group of molecules critical to guidance of the developing growth cone are proteoglycans and, again, calcium may mediate intracellular signalling by them. Snow et al. describe techniques for investigating calcium and other changes during neuronal growth. A thorough review of axon-growth inhibitory molecules is provided by Shewan and Cohen, who emphasize the need to characterize receptors and signalling systems for these inhibitory ligands. The final two chapters switch to a discussion of the role of electrical fields in axonal growth and guidance, and provide examples of the phenomenon and model experimental systems for evaluating these effects.

Both the volume editor and the series editors would like to thank all the authors for their timely contributions to our understanding of the neuronal growth cone, a real 'frontier' in neurobiology.

A.J. Turner
Leeds, 1996

Contributors

P. B. Atkinson
Colorado State University, Department of Anatomy and Neurobiology, Ft. Collins, CO 80523, U.S.A.

T. J. Baldwin
Department of Experimental Pathology, UMDS Guy's Hospital, London Bridge, London SE1 9RT, U.K.

C. E. Bandtlow
Brain Research Institute, University of Zurich, August-Forel-Str. 1, 8029 Zurich, Switzerland

M. S. Bush
The Randall Institute, King's College London, 26–29 Drury Lane, London WC2B 5RL, U.K.

J. Cohen
Department of Developmental Neurobiology, UMDS, Guy's Campus, London Bridge, London SE1 9RT, U.K.

R. W. Davenport
Laboratory of Developmental Neurobiology, National Institute of Child Health and Human Development, National Institutes of Health, Bethesda, MD 20892, U.S.A.

P. Doherty
Department of Experimental Pathology, UMDS Guy's Hospital, London Bridge, London SE1 9RT, U.K.

L. Erskine
Department of Anatomy & Developmental Biology, University College London, Windeyer Building, Cleveland Street, London W1P 6DB, U.K.

D. J. Goldberg
Department of Pharmacology and Center for Neurobiology and Behaviour, Columbia University, 630 W. 168th St., New York, NY 10032, U.S.A.

P. R. Gordon-Weeks
The Randall Institute, King's College London, 26–29 Drury Lane, London WC2B 5RL, U.K.

P. Grabham

Department of Pharmacology and Center for Neurobiology and Behaviour, Columbia University, 630 W. 168th St., New York, NY 10032, U.S.A.

T. D. Hassinger

Department of Neurobiology and Anatomy, The University of Utah, Salt Lake City, UT 84132, U.S.A.

M. Johnstone

The Randall Institute, King's College London, 26–29 Drury Lane, London WC2B 5RL, U.K.

S. B. Kater

Department of Neurobiology and Anatomy, The University of Utah, Salt Lake City, UT 84132, U.S.A.

P. C. Letourneau

University of Minnesota, Department of Cell Biology and Neuroanatomy, 321 Church St., SE, 4-135 Jackson Hall, Minneapolis, MN 55455, U.S.A.

C. D. McCaig

Department of Biomedical Sciences, Marischal College, University of Aberdeen, Aberdeen AB9 1AS, Scotland

M. A. Messerli

Department of Biological Sciences, Purdue University, West Lafayette, IN 47907, U.S.A.

K. R. Robinson

Department of Biological Sciences, Purdue University, West Lafayette, IN 47907, U.S.A.

D. Shewan

Department of Developmental Neurobiology, UMDS, Guy's Campus, London Bridge, London SE1 9RT, U.K.

D. M. Snow

University of Minnesota, Department of Cell Biology and Neuroanatomy, 321 Church St., SE, 4-135 Jackson Hall, Minneapolis, MN 55455, U.S.A.

F. S. Walsh

Department of Experimental Pathology, UMDS Guy's Hospital, London Bridge, London SE1 9RT, U.K.

D. -Y. Wu

Department of Pharmacology and Center for Neurobiology and Behaviour, Columbia University, 630 W. 168th St., New York, NY 10032, U.S.A.

Abbreviations

ACh	acetylcholine
AChR	acetylcholine receptor
AM	acetoxymethyl ester
BDNF	brain-derived neurotropic factor
bFGF	basic fibroblast growth factor
BNC-PG	bovine nasal cartilage proteoglycan
$[Ca^{2+}]_i$	intracellular Ca^{2+} concentration
CaM	calmodulin
CAM	cell adhesion molecule
CamK	calcium calmodulin-dependent protein kinase
CCD	charge-coupled device
C-domain	central domain
CHD	CAM homology domain
CKII	casein kinase II
CNS	central nervous system
Con A	concanavalin A
CPA	cyclopiazonic acid
CS	chondroitin sulphate
C-4-S	chondroitin-4-sulphate
C-6-S	chondroitin-6-sulphate
CSPG	chondroitin sulphate proteoglycan
DAG	diacylglycerol
DC	direct current
DRG	dorsal root ganglion
E	embryonic day
ECM	extracellular matrix
ELF-1	*Eph* ligand family-1
ERK	extracellular signal-related kinase
ERM	ezrin/radixin/moesin
F-actin	filamentous actin
FGF	fibroblast growth factor
FGFR1	FGF receptor 1
FITC	fluorescein isothiocyanate

GAG	glycosaminoglycan
GPI	glycosylphosphatidylinositol
HMM	high molecular mass
HSPG	heperan sulphate proteoglycan
InsP_3	inositol-1,4,5-trisphosphate
MAG	myelin-associated glycoprotein
MAP	microtubule-associated protein
MAP 1B-P	phosphorylated MAP 1B
MAP kinase	mitogen-activated protein kinase
NCAM	neural cell adhesion molecule
NGF	nerve growth factor
NF-H	heavy neurofilament proteins
PC12 cell	pheochromocytoma cell
PDPK	proline-directed protein kinase
PG	proteoglycan
PKC	protein kinase C
PLC	phospholipase C
PLCγ	phospholipase Cγ
p-membrane	E10 chick posterior tectum membrane
PNA	peanut agglutinin
PNS	peripheral nervous system
PP	protein phosphatase
P-region	peripheral region
PSA	polysialic acid
PTK	protein tyrosine kinase
RAGS	repulsive axon guidance signal
RC-PG	rat chondrosarcoma proteoglycan
RGC	retinal ganglion cell
RITC	rhodamine isothiocyanate
SCG	superior cervical ganglion
tau$_J$	juvenile tau
TEP	transepithelial potential
TG	thapsigargin
Δ^9-THC	Δ^9-tetrahydrocannabinol
TNTP	transneural tube potential
TrkA	NFG-receptor PTK
VDCC	voltage-dependent calcium channel

Overview: axon growth and axon guidance

C.D. McCaig

Department of Biomedical Sciences, Marischal College, University of
Aberdeen, Aberdeen AB9 1AS, Scotland, U.K.

Introduction

The following chapters grew out of a 1-day symposium held during the
Quincentenary of the University of Aberdeen. Two predominant issues dictated the
planning of the contributions. First, there would be no possibility of covering such
a large subject in a single day. The one comprehensive treatment of the field grew
from a 6-day meeting to celebrate 100 years of the nerve growth cone and involves
almost 40 chapters and more than 500 pages. Although this is now 5 years old, it
remains a marvellous sourcebook [1]. Selectivity was essential therefore, while
trying to retain some breadth of coverage. Secondly, given my own interest in
endogenous electric fields and their effects on nerve growth, this would be an
unashamed attempt to re-integrate galvanotropism, with 'mainstream' axon
guidance. Unashamed, because recent work indicates a role for endogenous electric
fields in embryonic development (Chapter 9) and because our own work on the
cellular mechanism of galvanotropism implicates key receptors and second
messenger systems, which are central to 'mainstream' axon guidance and which
may be shared by several guidance cues (see Chapter 10). The chapters which follow
therefore deal in some depth with receptor and second messenger mechanisms
involved in the transduction of extrinsic cues by the growth cone into guided
growth, before closing with two chapters which put the case for galvanotropism.

This overview will highlight some of the related areas of the field that are
not covered elsewhere. It will also interweave, where pertinent, aspects of electric-
field-guided nerve growth. In one sense therefore it might make easier reading after
the other chapters have been tackled.

The cytoskeleton and nerve growth and guidance

There is no specific coverage of issues relating to the supply by axonal transport of materials required for guided motility. While the role of microtubules in the growth cone is discussed (Chapter 4), the controversy relating to whether microtubules are transported *en bloc* to the growth cone is not. A new model for microtubule dynamics in a growing nerve shaft has been presented [2]. This incorporates both plus-end-leading microtubule transport *en bloc* and plus-end-specific microtubule assembly *en route*. One aspect of the model is that it takes account of the ability of nerve shafts to reorganize their cytoskeletons in response to extrinsic cues and to back-branch, thus creating a new sensory-motor growth cone out of what used to be thought of as simply a structural shaft. There is growing evidence that the neurite shaft may share some of the responsiveness of the growth cone to localized branching/guidance signals. Mammalian cortico-spinal projections are induced to back-branch by a pontine-derived factor which probably exists as a chemical gradient [3,4]. Back-branching of rat retinal axons also controls the retinotopic projection within the superior colliculus. In this case active suppression of branching by a membrane-bound inhibitory cue derived from the caudal superior colliculus is involved [5]. DC (direct current) electric fields also promote the sprouting of cathodally directed neurite branches [6–9]. Directed branching is predicated and predicted by an elevated intracellular Ca^{2+} concentration ($[Ca^{2+}]_i$) in the nerve shaft [9] and is inhibited by the nicotinic antagonist d-tubocurarine and by inhibitors of the inositol phospholipid second messenger system [7,8], indicating that common mechanisms may be involved in growth cone guidance and in the induction of site-directed branching. Given that focally applied acetylcholine (ACh) can induce growth cone guidance [10], it is possible that similar receptor-mediated events driven by focally elevated neurotransmitters or growth factors might control back-branching of neurites.

The intrinsic mechanisms underlying growth cone motility are not discussed overtly in any chapter. Recent work has built on the seminal publications of Mitchison and Kirschner [11] and of Smith [12], to provide sophisticated models of the cytoskeletal reorganization which underpins directed growth cone motility [13,14]. These incorporate details of the probable molecular linkages involved in the coupling of extracellular signals to directed cytoskeletal polymerization. A model is developed wherein substrates which promote growth do so by assembling a set of clutch proteins which couple the extracellular matrix ligand/cellular receptor complex to the retrograde flow of filamentous actin (F-actin). F-actin retrograde flow is slowed through this coupling of the cytoskeleton to the substrate and this functionally transduces the force-generating retrograde flux of F-actin into growth cone advance [13]. There is experimental support for this model from *Aplysia* bag cell neurons. Using polycationic beads as surface markers which move at the same rate as the underlying F-actin, and laser tweezers to place beads in specific growth cone locations during growth cone interactions with a target neuron, Lin and

Forscher [15] have shown elegantly that retrograde flow of F-actin is attenuated specifically along the axis of target interaction and that this happens in direct proportion to the rate of growth cone advance. Target site-directed microtubule extension and stabilization depends also on the accumulation of F-actin at a site of target contact [16]. Clearly, this model ascribes a key role to F-actin dynamics in steering the growth cone. Whether similar cytoskeletal events underpin site-directed back-branching of neurites has still to be examined.

Genetics and axon growth and guidance

None of the following chapters deals with the genetics of axon growth and guidance. There is good coverage, however, of spatially restricted regulatory gene expression in the developing neuroepithelium, which determines the laying down of an inherent patterning to direct pioneering growth cones [17]. The genetic basis of selective fasciculation as a mechanism for bundling and guiding related axons also has been identified in *Drosophila*. Loss of function mutations of the gene *fasc II* which encodes a *Drosophila* neural cell adhesion molecule (NCAM) results in defective neuronal bundling and guidance [18]. Recent genetic evidence also implicates receptor protein tyrosine kinase (PTK) involvement in neuronal pathfinding. The *Drosophila* gene *derailed* encodes a novel member of the receptor PTK family and is expressed by a small subset of embryonic interneurons with growth cones that follow common pathways. In *drl* mutant embryos, these specific neurons fail to pathfind appropriately [19]. Receptor PTKs are being identified increasingly as key elements in signalling aspects of neuronal growth and guidance (see Chapters 2 and 3). This is one signalling element which electric fields may share with other guidance mechanisms. We have preliminary evidence that the receptor PTK inhibitor RG50864, which prevents electric field-induced redistribution of ACh receptors on *Xenopus* myoblasts [20], also prevents cathodal-directed turning of *Xenopus* embryonic spinal neurites [21].

Chemoattraction and chemorepulsion: could gradients be established by endogenous electric fields?

Recent genetic work on nerve guidance in the nematode *Caenorhabditis elegans* provides data of interest in cross comparison with the recently discovered chemo-attractant and chemorepellant mammalian netrins. The *unc 5* gene encodes a transmembrane receptor expressed on the surface of migrating cells and growth cones [22]. Unc 6 is a secreted protein present in basal lamina which is structurally related to laminin and which may be an adhesive ligand that activates the Unc 5 receptor. Both proteins are required for circumferential migration of some cells and of pioneer axons in the nematode and may form a guidance mechanism controlling these migrations. Importantly, it has been suggested that a gradient of secreted Unc 6 may attract some axons, while repelling others [22]. The netrins are homologues of Unc 6. Netrins 1 and 2 act as secreted, long-range chemoattractants for developing spinal commissural axons *in vitro*, while netrin 1 is expressed by cells in the floor plate which is an intermediate target of commissural axon growth [23]. Moreover, there is direct evidence that netrin 1 is a bifunctional guidance cue (as suggested for Unc 6), in that it acts as a diffusible chemorepellant for a separate set of axons, i.e. trochlear motor axons [24]. This emerging concept of bi- or multifunctionality of guidance cues is intriguing and applies equally well to endogenous DC electric fields (see Chapter 10). *In vitro*, a DC field may have different effects on different cell types, but also may selectively attract or repel oppositely directed neurites on the same cell body. Additionally, the ability to direct back-branching establishes DC fields as potent multifunctional guidance cues.

Netrin 1 is the most recent addition to a growing group of disparate inhibitory cues, which may exist as substrate-bound molecules or in the form of a diffusible chemical gradient (see Chapter 8 and [5,25–27]). Many of these, netrin 1, Unc 6, collapsin and other members of the semaphorin family of secreted proteins are highly positively charged proteins. Given the evidence presented by Robinson and Messerli (Chapter 9) showing that DC electric fields exist in association with the developing neural tube and that standing gradients of charged proteins can be established *in vitro* by fields of similar strength, perhaps endogenous fields play a role in establishing the gradients of these chemorepellant and chemoattractant molecules.

Chemoaffinity

Finally, one area touched on by Shewan and Cohen (Chapter 8) has recently seen major progress. Sperry's proposal [28] that precise retino-tectal connections are governed by sets of complimentary cytochemical tags, is now supported by identified receptors and ligands arrayed in complimentary gradients [29–32]. Two membrane-bound glycosylphosphatidylinositol-linked proteins have been identified as potential ligands expressed in a smooth gradient across the embryonic chick tectum. RAGS (repulsive axon guidance signal) [29] and ELF-1 (Eph ligand family-1) [30] are each expressed at high levels in posterior chick tectum and in a diminishing gradient down to low levels in anterior tectum. While the receptor for RAGS is unclear, ELF-1 binds with high affinity to the Mek-4 receptor, which is a member of the Eph family of receptor tyrosine kinases. Mek-4 receptors also were found to be expressed in a gradient across the retina, such that temporal retinal ganglion cells were labelled more strongly than nasal retinal ganglion cells [30]. ELF-1 and Mek-4 are expressed at the right time and with appropriately opposing gradients to be part of a chemoaffinity mechanism for mapping retinal axons topographically on to the tectum.

A recent short review of axon guidance molecules provides an appropriate end point. Baier and Bonhoeffer [33] present an attractive Figure in which six different types of molecule are shown to have differing effects on axon guidance, depending on whether they are presented as a homogeneous distribution or as a concentration gradient. The groups of molecules are permissive (e.g. laminin), inhibitory (e.g. NI-35), outgrowth-promoting (e.g. netrins), outgrowth-suppressing (e.g. chemorepulsive activities [34] or netrin 1 for trochlear motoneurons [24]); attractive (e.g. netrins/nerve growth factor); and repulsive (e.g. chemorepulsive activities [34] or netrin 1 for trochlear motoneurons [24]). In the current spirit of bi- or multi-functional guidance cues, one cue, a DC electric field, is capable of all of these activities *in vitro* (see Chapter 10). Determining that such a plethora of activities occurs also *in vivo* would project electric fields from the poor relation of axon guidance to one of the aristocrats. Such work has begun and is promising (see Chapter 9).

I thank The Wellcome Trust for support.

References

1. Letourneau, P.C., Kater, S.B. and Macagno, E.R., eds. (1991) The Nerve Growth Cone, Raven Press, New York
2. Joshi, H.C. and Baas, P.W. (1993) J. Cell Biol. **121**, 1191–1196
3. Heffner, C.D., Lumsden, A.G.S. and O'Leary, D.D.S. (1990) Science **247**, 217–220
4. Sato, M., Lopez-Mascaraque, L., Heffner, C.D. and O'Leary, D.D.S. (1994) Neuron **13**, 791–803
5. Roskies, A.L. and O'Leary, D.D.S. (1994) Science **265**, 799–803
6. McCaig, C.D. (1990) J. Cell Sci. **95**, 605–615
7. Erskine, L., Stewart, R.S. and McCaig, C.D. (1995) J. Neurobiol. **26**, 523–536
8. Erskine, L. and McCaig, C.D. (1995) Dev. Biol. **171**, 330–339
9. Williams, C.V., Davenport, R.W., Dou, P. and Kater, S.B. (1995) J. Neurobiol. **27**, 127–140
10. Zheng, J.Q., Felder, M., Connor, J.A. and Poo, M.-M. (1994) Nature (London) **368**, 140–144
11. Mitchison, T. and Kirschner, M. (1988) Neuron **1**, 761–772
12. Smith, S.J. (1988) Science **242**, 708–715
13. Lin, C.-H., Thompson, C.A. and Forscher, P. (1994) Curr. Opin. Neurobiol. **4**, 640–647
14. Bentley, D. and O'Connor, T.P. (1994) Curr. Opin. Neurobiol. **4**, 43–48
15. Lin, C.-H. and Forscher, P. (1995) Neuron **14**, 763–771
16. Lin, C.-H. and Forscher, P. (1993) J. Cell Biol. **121**, 1369–1383
17. Wilson, S.W., Placzek, M. and Furley, A.J. (1993) Trends Neurosci. **16**, 316–323
18. Lin, D.M., Fetter, R.D., Kopczynski, C., Grenninggoh, G. and Goodman, C.S. (1994) Neuron **13**, 1055–1069
19. Callahan, C.A., Muralidhar, M.G., Lundgren, S.E., Scully, A.L. and Thomas, J.B. (1995) Nature (London) **376**, 171–174
20. Peng, H.B., Baker, L.P. and Dai, Z. (1993) J. Cell Biol. **120**, 197–204
21. Allan, D.W. (1994) M.Sc Thesis, University of Aberdeen, Aberdeen
22. Ishii, N., Wadsworth, W.G., Stern, B.D., Culotti, J.G. and Hedgecock, E.M. (1992) Neuron **9**, 873–881
23. Kennedy, T.E., Serafini, T., de la Torre, J.R. and Tessier-Lavigne, M. (1994) Cell **78**, 425–435
24. Colomarino, S.A. and Tessier-Lavigne, M. (1995) Cell **81**, 621–629
25. Keynes, R.J. and Cook, G.M.W. (1995) Curr. Opin. Neurobiol. **5**, 75–82
26. Schwab, M.E., Kapfhammer, J.P. and Bandtlow, C.E. (1993) Annu. Rev. Neurosci. **16**, 565–595
27. Dodd, J. and Schuchardt, A. (1995) Cell **81**, 471–474
28. Sperry, R.W. (1944) J. Neurophysiol. **7**, 57–69

29. Drescher, U., Kremoser, K., Handwerker, C., Löschinger, J., Noda, M. and Bonhoeffer, F. (1995) Cell **82**, 359–370
30. Cheng, H.-J., Nakamoto, M., Bergemann, A.D. and Flanagan, J.G. (1995) Cell **82**, 371–381
31. Harris, W.A. and Holt, C.E. (1995) Neuron **15**, 241–244
32. Tessier-Lavigne, M. (1995) Cell **82**, 345–348
33. Baier, H. and Bonhoeffer, F. (1994) Science **265**, 1541–1542
34. Pini, A. (1993) Science **261**, 95–98

2

Protein tyrosine phosphorylation in the growth cone

Peter W. Grabham, Da-Yu Wu and Daniel J. Goldberg

Department of Pharmacology and Center for Neurobiology and
Behaviour, Columbia University, 630 W. 168th St., New York,
NY 10032, U.S.A.

Introduction

The nerve growth cone is a remarkable structure. It is able to detect environmental
cues, transduce them into signals for the direction and rate of axonal growth, and
then mediate that growth by altering its morphology and motility. Crucial to this
process is the peripheral region (P-region) of the growth cone which is comprised
of the lamellipodium (veils of membrane) and the major sensory elements of the
growth cone, the digitate filopodium (Fig. 1A). Video-microscopic studies of living
growth cones show that the P-region is highly motile [1]. The distal margin
continually remodels as filopodia and veils rapidly extend, retract and wave about.
Moreover, there is a continual flow of protrusions on the surface of the
lamellipodium towards the rear.

These motile activities have been functionally correlated to axonal
growth. For example, it has been observed of certain axons growing *in vivo* that the
growth cone is streamlined and simple when the neurite is growing along a straight
path, but generates a large lamellipodium, numerous filopodia and is vigorously
active when it pauses in a decision point where a choice between alternative paths
must be made [2,3]. The interpretation of such observations is that growth cones at
decision points are more elaborate and more motile in order to effectively survey
the environment for the next cue. Furthermore, when an axon eventually reaches its
synaptic target, motile activity must shut down to convert it into a synaptic
terminal. Thus, the plasticity and motility of the growth cone mediates its function
which, in turn, mediates axonal growth and synapse formation.

The mechanisms that control these events in the growth cone are poorly
understood. How does an environmental cue such as an extracellular matrix (ECM)
protein, a trophic factor, or a surface glycoprotein rapidly control the activity of the
growth cone? Several lines of evidence implicate protein tyrosine phosphorylation
in local growth cone regulation. First, a number of cytoskeleton-associated

molecules present in growth cones (vinculin, integrin, ezrin, cortactin) are tyrosine phosphorylated, and tyrosine phosphorylation represents a rapid means of altering protein function. Secondly, neurotrophic factors such as nerve growth factor (NGF) and fibroblast growth factor bind to receptors which are protein tyrosine kinases (PTKs) [4–6] and binding of these ligands not only causes autophosphorylation of the receptors but tyrosine phosphorylation of numerous other proteins as well [7–9]. Thirdly, non-receptor PTKs have also been implicated in growth cone function. Non-receptor PTKs of the *src* family (pp60^{c-src}, pp62^{c-yes} and p59^{fyn}) are expressed at high levels in brain, particularly in growth cones, pp60^{src} levels peak during the time of axonal growth, and there is a neuronal specific form of pp60^{src} that is concentrated in growth cones [10–14]. Another PTK, *abl*, is transiently expressed in certain axonal tracts of the developing *Drosophila* nervous system and genetic deletion of *abl* contributes to wayward axonal growth [15]. Finally, experiments using mutant mice with deletions for pp60^{src}, pp62^{yes} and p59^{fyn} have demonstrated that certain of these non-receptor tyrosine kinases are components of the intracellular signalling pathway mediating neurite outgrowth on molecules of the immunoglobulin superfamily. Specifically, src is required for maximal L1-mediated axonal outgrowth [16], and fyn is required for maximal NCAM (neural cell adhesion molecule)-mediated axonal outgrowth [17].

The actions of neurotrophic factors are well known to involve classic second messengers that include not only receptor PTKs but other tyrosine phosphorylation events downstream of the receptor [4]. What are unknown, however, are specific roles for tyrosine phosphorylation in eliciting changes in growth cone morphology and motility, and consequently, the rate and pathfinding of axonal growth. Furthermore, it is clear that there are multiple tyrosine phosphorylation events regulating axonal growth. This is evident from a consideration of pharmacological studies using tyrosine kinase inhibitors. Such experiments appear superficially contradictory; tyrosine kinase inhibitors have been shown to both inhibit [18–20] and stimulate [21,22] neurite outgrowth. Although the targets for different pharmaceutical agents may differ, a probable reason for these contradictions is the existence of a complex array of PTK activities that act in concert and/or opposition to regulate several processes leading to axonal outgrowth. While it is clear that the inhibitory effects of some PTK inhibitors are due to their action on the NGF-receptor PTK (TrkA) at the start of the signalling cascade [18–20], later events are not well defined. For example, it has been shown in cerebellar granule cells that neurite outgrowth is stimulated by only one PTK inhibitor (herbimycin A). Furthermore, this stimulation is blocked by two other tyrosine kinase inhibitors that on their own do not affect neurite outgrowth [23].

The subject of our studies is not the transcriptionally dependent onset of neuronal differentiation and the initial stimulation of axon formation, but the acute effects of environmental cues on tyrosine phosphorylation in the growth cone itself. How do these cues act locally and rapidly at the growth cone to alter morphology and motility, and therefore, the pathfinding and rate of axonal growth? We describe here two tyrosine phosphorylation events in the growth cone that may be involved in growth cone function: first, the triggering and sustenance of motility of the growth cone and secondly, the functioning of the sensory element of the growth cone, the filopodium.

Fig. 1.   **Video-micrographs of an *Aplysia* growth cone on a polylysine substrate**

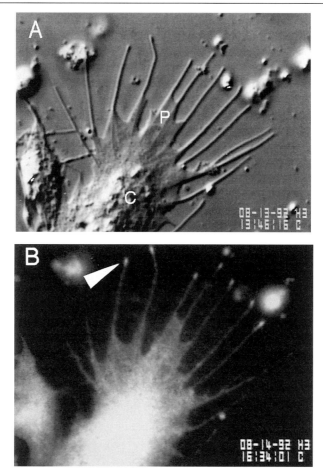

(A) VEC-DIC (video-enhanced contrast–differential-interference contrast) microscopy shows a central region (C) containing membrane-bound organelles and microtubules, and an actin-rich peripheral region (P) with lamellipodia and digitate filopodia. (B) Staining with anti-phosphotyrosine primary antibody and fluorescein-conjugated secondary antibody shows that phosphotyrosine is concentrated at the distal ends of the filopodia (indicated by arrow).

Generation of motility in the growth cone

As mentioned above, motility in the growth cone is highly variable. Activity is greatest when the axon pauses at 'decision points' and much less when the axon is rapidly extending. Any study of motility requires an experimental model in which this activity can be selectively stimulated. This raises the question of which environmental cues stimulate motility as opposed to extension. This is a difficult question to answer using experimental models, since more than one cue is generally required to induce neurite outgrowth *in vitro*. For example, in rat pheochromocytoma cells (PC12 cells), differentiation to neurite-bearing sympathetic-like neurons requires the continued presence of both the neurotrophic factor (NGF) and an ECM molecule such as collagen or laminin [24]. Furthermore, many primary neuronal cultures require non-defined mixtures of trophic factors (e.g. serum in vertebrates or haemolymph in molluscs) in addition to an ECM molecule for survival and axonal outgrowth. It is likely that both extension cues and motility cues act in a delicate balance to produce functional growth cones and, consequently, axonal outgrowth. In spite of this difficulty, it is possible to manipulate culture conditions to induce motile activity. Here we will describe two experimental models in which we have examined tyrosine phosphorylation events following the generation of motile activity in the growth cone.

Aplysia axotomy

Neurons of the mollusc *Aplysia* are large and can be removed from the animal and placed in culture with a substantial length of the original axon still attached to the cell body. Like the axons of other neurons, they contain three major classes of cytoskeletal filaments: microtubules, neurofilaments and actin microfilaments. The distribution of these filaments provides clues to the mechanisms that control growth cone motility. Both neurofilaments and microtubules are major longitudinal structural components of the developing axon. Neurofilaments rarely extend into the growth cone, whereas microtubules often extend into the central region of the growth cone and can sometimes be seen terminating near the leading edge of the P-region. Actin is found as bundles of parallel filaments that form the core of filopodia, and in a meshwork within the lamellipodium [25,26] (Fig. 2). Thus, the major cytoskeletal filament in the motile P-region of the growth cone is actin and changes in motility in this region probably involve a rapid and directed reorganization of the actin network. Two possible mechanisms to account for this reorganization are: (1) a shift in the balance between polymerization and depolymerization; and (2) the movement of actin filaments, which is likely to be driven by actin–myosin interactions.

Motile protrusions can be stimulated in these neurons by axotomy [27]. Transection of the axon rapidly transforms quiescent cytoplasm into motile cytoplasm, with large numbers of filopodia and expanses of veil rapidly forming as the first manifestations of the motile response (Fig. 3). The trigger for such activity is unknown, but we hypothesize that it involves tyrosine phosphorylation of actin-associated proteins. Staining with an antibody which specifically recognizes phosphorylated tyrosine residues suggests that PTK activity is involved because phosphotyrosine concentrates in the areas from which protrusions are beginning to emerge (Fig. 4). Furthermore, studies using a variety of kinase inhibitors reveals that only inhibitors of PTKs block the formation of protrusions. Since the same PTK inhibitors cause disruption of the actin network of the lamellipodium when applied to existing growth cones, it is likely that they are interfering with the assembly of actin filament networks during the generation of motile protrusions. It should be interesting to determine by immunostaining whether the phosphotyrosine content is elevated in growth cones *in vivo*, particularly when they have paused at 'decision points' and are starting to send out protrusions. In addition, immunostaining of transected neurons with antibodies raised against proteins which are known to be both actin-associated and tyrosine-phosphorylated, might provide clues to the mechanism(s) involved in growth cone motility.

Fig. 2.   **Actin distribution in an *Aplysia* growth cone**

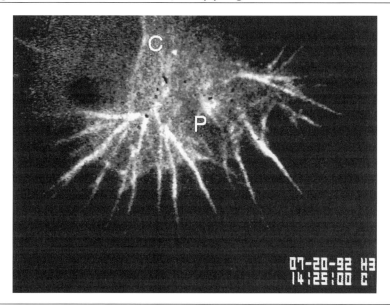

Staining with fluorescent phalloidin which visualizes filamentous actin shows a concentration in the peripheral region (P) and the core bundles of filopodia.

Fig. 3.   Generation of motile activity following axotomy of an *Aplysia* axon

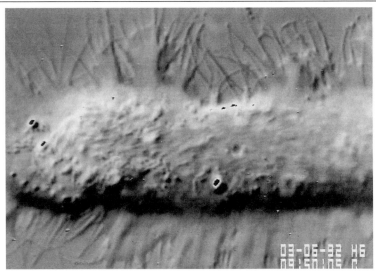

VEC-DIC microscopy 20 min after axotomy, illustrates extensive motile activity typified by sheets of lamellipodia and numerous filopodia. Motile activity can be seen both at the newly cut tip and for some distance along the axon shaft.

Fig. 4.   Aggregates of phosphotyrosine appear at the margins of an *Aplysia* axon 5 min after axotomy

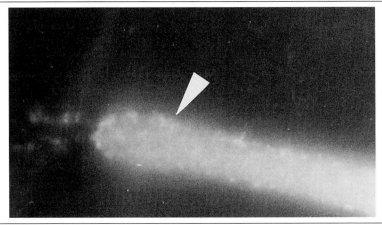

Staining with anti-phosphotyrosine primary antibody and fluorescein-conjugated secondary antibody, visualizes aggregates of phosphotyrosine (arrowhead) which have appeared at the margins of the axon.

Fig. 5. **Rapid NGF-induced tyrosine phosphorylation in neurite fractions of primed PC12 cells**

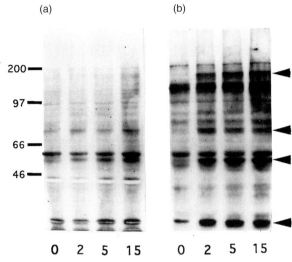

NGF (50 ng/ml) was added to cells for 2, 5 and 15 min as indicated. Neurite fractions were separated into a cytoskeletal fraction (**a**) and a Triton-X-soluble fraction (**b**), then immunoblotted using an antibody to phosphotyrosine. Molecular-mass markers are shown on the left-hand side. Proteins that increase their phosphotyrosine content are indicated by arrowheads on the right-hand side.

NGF withdrawal/readdition paradigm

Although *Aplysia* neurons offer the advantage of size for microscopy and immuno-fluorescence studies, it is difficult to obtain enough material for biochemical or molecular studies. We have therefore used a protocol of addition of NGF to the PC12 cell line to study biochemical events during the activation of growth cone motility. This model was first described by Greene and co-workers [28–30]. PC12 cells on a collagen substrate are differentiated by treatment with NGF for 6–8 days. The resulting neurites are tipped by growth cones with a remarkable variety of shapes and activities, ranging from simple, quiescent club-shaped endings to elaborate, highly motile endings. NGF is then removed for a period of 7 h or more by washing with NGF-free medium. After withdrawal, nearly all neurites are blunt-ended and, except for a few filopodia, are not motile; neurite elongation is stopped. When NGF is added again, a rapid increase in growth cone activity is observed. Within 2 min, numerous filopodia appear at the tips and, in some cases, along the shaft of the neurite. This is followed by a progressive extension of protrusions (lamellipodia, ruffles, filopodia) until, within 20 min, the majority of growth cones display a highly motile P-region. At 20–30 min after re-addition, neurites begin to elongate. The speed at which motility in the P-region occurs indicates that post-translational modifications are likely to be involved and that they are under local control, without communication from the cell body. This is confirmed by experiments where neurites detached from the cell body have been shown to display a similar rapid onset of growth cone motility following the addition of NGF [29]. In addition, the observation that neurite extension starts at least 15 min after the onset of motility lends support to the notion that these two processes are distinct.

We have used a protocol adapted from that of Sobue and Kanda [31], for the fractionation of 'growth cone-enriched' neurites, to examine tyrosine phosphorylation changes associated with the onset of motility. Briefly, motility is induced as described above, cell bodies are separated from neurites by trituration followed by centrifugation over a 20% (w/v) sucrose gradient, and the neurite fraction pelleted at 100 000 g. This is then further fractionated by detergent [1% (v/v) Triton-X] extraction to yield enriched fractions of detergent-soluble proteins and, of particular interest here, cytoskeletal elements plus associated proteins. Proteins are resolved by SDS/PAGE and immunoblotted using an antibody directed against tyrosine-phosphorylated proteins. Since these cells are already primed, changes in protein tyrosine phosphorylation reflect either the early stages of the NGF signalling pathway or events associated with the onset of motility, and not long-term changes seen during priming of PC12 cells. It can be seen in Fig. 5 that at least four protein species increase their phosphotyrosine content at a time (within 2 min) that corresponds to the onset of motility. These are, however, mostly in the detergent-soluble fraction. Immunoprecipitation by the same antibody coupled to Sepharose reveals a similar pattern of tyrosine phosphorylation, indicating that binding is specific. A protein at 140 kDa probably represents the TrkA receptor, which is well known to autophosphorylate upon NGF binding [32]. The 34 kDa species has previously been observed to become phosphorylated solely by activation of TrkA [33] and is thought to represent one of the earliest biochemical markers involved in a unique signalling pathway for NGF. It is, however, not yet clear whether this protein is a proteolytic fragment of TrkA or not. A major band at 55–60 kDa can be seen to become rapidly (within 2 min) phosphorylated in the Triton-X-soluble fraction and also somewhat later (15 min) in the cytoskeleton fraction. This protein is probably α-tubulin since it co-migrates with α-tubulin and since α-tubulin is known to become rapidly tyrosine-phosphorylated following NGF treatment [17]. The extracellular signal-related kinases (ERKs), ERK 1 and ERK 2, which are known to be tyrosine-phosphorylated after NGF treatment [34,35], are detected in both neurite and cell-body fractions using a polyclonal antibody directed against ERKs. However, the tyrosine-phosphorylated forms of these proteins were only observed in cell-body fractions. A protein at around 75 kDa is also seen to become rapidly tyrosine-phosphorylated in Triton-X-soluble neurite fractions. We have not, to date, been able to identify this protein but it is of interest because it is specific to neurites.

In conclusion, we have not detected cytoskeletal proteins (other than α-tubulin) which consistently change their phosphotyrosine content following the onset of growth cone motility. This may be a sensitivity problem because the relative volume of the growth cone is small compared with the rest of the neurite. As a further purification step we are currently isolating actin-associated proteins using immobilized actin bound to fractionation columns. In addition, we are attempting to determine whether the 75 kDa tyrosine-phosphorylated protein is an actin-binding protein by using a gel overlay technique.

Signal detection by filopodia

The digitate filopodia that extend tens of micrometers from the body of the growth cone are key sites for interaction with the environment. Contact of a single filopodium with positive [36] or negative [37,38] cues can be sufficient to reorient an axon. For example, when a filopodium of a Ti1 pioneer neuron in the developing grasshopper nervous system touches a guidepost cell, it will form the basis of a new direction of growth towards the guidepost cell [39–41]. As organelles move forward and fill the filopodium, other filopodia and branches wither. Suppression of the formation of filopodia results in wayward growth [42].

The mechanism(s) by which filopodia act as sensory structures and mediate reorientation are largely unknown. Recent observations in our laboratory suggest a role for tyrosine phosphorylation in filopodial function. When immuno-stained with an antibody directed against phosphotyrosine, *Aplysia* growth cones of axons growing slowly on polylysine in a protein-free environment show a diffuse staining pattern. There is, however, an intense concentration of phosphotyrosine staining at the filopodial tips [43] (Fig. 1B). This phenomenon is not restricted to *Aplysia* since tip staining has also been observed in a variety of vertebrate neurons in culture including embryonic chick sympathetic and retinal neurons, embryonic mouse retinal ganglion cells, and neonatal rat sympathetic and hippocampal neurons.

The environment can influence phosphotyrosine at filopodial tips. When laminin is added to cultures of mouse retinal ganglion cells, an increase in the rate of neurite extension is coincident with a reduction in the number of tips with high concentrations of phosphotyrosine [44]. The evidence that tip phosphorylation can be regulated by the environment supports the notion that it is involved in mediating or facilitating interactions of the filopodium with the environment. This is further supported by evidence of a close association between tip phosphorylation and a receptor for ECM molecules. β1 Integrin, a subunit of the receptor that mediates the promotion of neurite growth by laminin, collagen and fibronectin, co-concen-trates with phosphotyrosine in filopodial tips both in *Aplysia* and vertebrates [44]. Moreover, the relationship between β1 integrin and phosphotyrosine is an intimate one, since analysis of *Aplysia* growth cones shows that virtually all filopodia that have aggregates of β1 integrin at their tips also have concentrated phosphotyrosine. This notion is further supported by the finding that pharmacological agents which cause phosphotyrosine to disappear from the tips also cause β1 integrin to disappear. Inhibition of PTK activity by genistein (a broad specificity PTK inhibitor), causes both to disappear. Disruption of actin by cytochalasin D, also causes both to disappear. As actin filaments are disrupted, phosphotyrosine withdraws distoproximally along the filopodium [43], and β1 integrin can occasionally be seen to maintain its co-localization with phosphotyrosine during this withdrawal. Furthermore, when actin filaments reappear after removal of cytochalasin, phosphotyrosine and β1 integrin return to filopodial tips simultaneously. These results suggest that the interaction of growth cone filopodia with ECM proteins is either facilitated or mediated by tyrosine phosphorylation. We will consider here two possible roles for tip phosphorylation in filopodial function.

Regulation of filopodial dynamics

It is possible that filopodial dynamics are regulated by tip phosphorylation. Filopodia are highly dynamic structures, constantly elongating and retracting. Underlying these dynamics is a constant backward flow of actin filaments from the tip towards the centre of the growth cone [45–47]. Equilibrium is maintained by the addition of subunits at the distal tip and disassembly in the central region of the growth cone. Because the uniformly polarized actin filaments are oriented with the kinetically active ('barbed') ends distal, addition of subunits is likely to occur at this site. Therefore, phosphotyrosine is strategically positioned to regulate the addition of actin subunits, and in turn, filopodial dynamics. Of importance here is the observation that dephosphorylation by the addition of PTK inhibitors results in filipodial lengthening [43]. Furthermore, addition of the ECM protein — laminin — also decreases tip phosphorylation [44], and although it eventually causes the growth cone to become streamlined, its immediate effect is a lengthening of filopodia by way of a change in dynamics similar to that seen with PTK inhibitors [43,48]. In this hypothesis, tip phosphorylation could be mediating an effect of an environmental cue on filopodial dynamics. For example, a cue which promotes rapid axonal growth could promote filopodial lengthening by causing tip dephosphorylation. A second example is the contact of a filopodium with a surface-bound cue with which a prolonged attachment might be required, such as a guidepost cell or a synaptic target. This could induce tip phosphorylation in order to suppress filopodial dynamics.

A possible means by which tyrosine phosphorylation might regulate actin polymerization, and therefore filopodial dynamics, is by phosphorylation of actin itself, as is the case in the slime mould *Dictyostelium* [49]. Alternatively, tyrosine phosphorylation could increase the affinity of another protein for the barbed end of the actin filament and thus block actin monomer addition. Several proteins are known to bind to barbed ends, and we have used immunofluorescence to determine whether any of these proteins co-concentrate with tyrosine at filopodial tips. To date, the only antibody which shows concentrated staining at filopodial tips [44] is one which recognizes members of the ezrin/radixin/moesin (ERM) family of proteins [50]. Furthermore, this antibody intensely stains most filopodial tips that have concentrated phosphotyrosine. Of the ERM proteins, radixin is known to bind to barbed ends [51], Whether its interaction with actin is influenced by phosphorylation, however, remains to be determined.

Although the notion that tyrosine phosphorylation modulates actin dynamics, and thus promotes filopodial lengthening, is an attractive hypothesis, evidence supporting it is scant. Furthermore, two other observations do not support this hypothesis. First, in active growth cones, there is no difference in dynamics between filopodia with phosphorylated tips and those without. It is interesting to note here, that while carrying out these observations we did notice that newly emerged filopodia do not have detectable tip concentrations; it takes minutes for phosphotyrosine to accumulate at the tip. Secondly, we have found in some *Aplysia* growth cones that a lower concentration of the PTK inhibitor genistein will, over the course of 2–3 h, deplete phosphotyrosine at filopodial tips without apparently causing substantial filopodial lengthening.

Localization of receptors

Experiments we have carried out with the PTK inhibitor genistein suggest that tip phosphorylation is required to aggregate β1 integrin and possibly other receptors at the filopodial tip. Treatment of *Aplysia* growth cones with low concentrations of genistein (50 μM) causes a disappearance of both the aggregates of β1 integrin and of phosphotyrosine from filopodial tips, without apparent effects on actin. There is evidence that tyrosine phosphorylation is involved in protein aggregation since it is known to promote the formation of protein complexes at activated growth factor receptors [52,53]. Alternatively, tyrosine phosphorylation might affect localization of a protein by modifying its ability to link to the cytoskeleton. For example, tyrosine phosphorylation of β1 integrin inhibits its linking to the actin cytoskeleton and reduces its localization to focal contacts [54–56].

Research has indicated that the tips of chick retinal neuron filopodia are particularly adhesive [57]. The molecular basis of this adhesiveness may be the concentration of integrin caused by tyrosine phosphorylation. In this case, the interactive relationship between the filopodium and the environment would be facilitated through tyrosine phosphorylation by fostering an accumulation of receptors. Thus, tyrosine phosphorylation would be increasing the sensitivity of filopodia to environmental cues.

Future analysis of the role(s) of tyrosine phosphorylation in filopodia should address several issues. First, which environmental cues are associated with changes in filopodial tyrosine phosphorylation. Also, what are the consequences of interaction with those cues? For example, is there an increase in tip phosphorylation when a filopodium contacts a guidepost cell but not when it transiently contacts irrelevant cells? Secondly, identification of both the substrate(s) and the kinase(s) for tyrosine phosphorylation in filopodial tips is critical for an understanding of the role of tip phosphorylation. ERM protein and β1 integrin, the two proteins we have so far localized to tip aggregates, are both substrates for tyrosine kinase activity. However, it is not yet known if these are indeed the phosphorylated proteins. In the case of the kinase(s), binding of ligand to integrin is known to recruit focal adhesion kinase (FAK) and, probably src kinases to focal contacts and may activate them [58,59]. We have not to date detected either kinase to the tip aggregates by immunocytochemistry, suggesting that they are not the candidate tyrosine kinase(s). Thirdly, are there receptors other than integrin associated with the phosphotyrosine? N-cadherin is known to be enriched at the cell–cell adherens junction which is also rich in phosphotyrosine [60]. It would be interesting to know if N-cadherin also co-concentrates with phosphotyrosine at filopodial tips.

Work in this laboratory was supported by the NIH.

References

1. Goldberg, D.J., Burmeister, D.W. and Rivas, R.J. (1991) in The Nerve Growth Cone (Letourneau, P.C., Kater, S.B. and Macagno, E.R., eds.), pp. 79–95, Raven Press, New York
2. Tosney, K.W. and Landmesser, L.T. (1985) J. Neurosci. **5**, 2345–2358
3. Bovolenta, P. and Mason, C. (1987) J. Neurosci. **7**, 1447–1460
4. Kaplan, D.R. and Stephens, R.M. (1994) J. Neurobiol. **25**, 1404–1417
5. Barbacid, M. (1994) J. Neurobiol. **25**, 1386–1403
6. Van der Geer, P., Hunter, T. and Lindberg, R.A. (1994) Annu. Rev. Cell Biol. **10**, 251–337
7. Maher, P.A. (1988) Proc. Natl. Acad. Sci. U.S.A. **85**, 6788–6791
8. Schanen-King, C., Nel, A., Williams, L.K. and Landreth, G. (1991) Neuron **6**, 915–922
9. Vetter, M.L., Martin-Zanca, D., Parada, L.F., Bishop, J.M. and Kaplan, D.R. (1991) Proc. Natl. Acad. Sci. U.S.A. **88**, 5650–5654
10. Cotton, P.C. and Brugge, J.S. (1983) Mol. Cell. Biol. **3**, 1157–1162
11. Brugge, J.S., Cotton, P.C., Queral, A.E. et al. (1985) Nature (London) **316**, 554–557
12. Ingraham, C.A., Cooke, M.P., Chuang, Y.N., Perlmutter, R.M. and Maness, P.F. (1992) Oncogene **7**, 95–100
13. Sudol, M., Alvarez-Buylla, A. and Hanafusa, H. (1988) Oncol. Res. **2**, 345–355
14. Bixby, J.L. and Jhabvala, P. (1993) J. Neurosci. **13**, 3421–3432
15. Elkins, T., Zinn, K., McAllister, L., Hoffmann, F.M. and Goodman, C.S. (1990) Cell **60**, 565–575
16. Ignelzi, M.A., Miller, D.R., Soriano, P. and Maness, P.F. (1994) Neuron **12**, 873–884
17. Beggs, H.E., Soriano, P. and Maness, P.F. (1994) J. Cell Biol. **127**, 825–833
18. Tsukada, Y., Chiba, K., Yamazaki, M. and Mohri, T. (1994) Biol. Pharm. Bull. **17**, 370–375
19. Ohmichi, M., Pang, L., Ribon, V. et al. (1993) Biochemistry **32**, 4650–4658
20. Muroya, K., Hashimoto, Y., Hattori, S. and Nakamura, S. (1992) Biochim. Biophys. Acta **1135**, 353–356
21. Bixby, J.L. and Jhabvala, P. (1992) J. Neurobiol. **23**, 468–480
22. Miller, D.R., Lee, G.M. and Maness, P.F. (1993) J. Neurochem. **60**, 2134–2144
23. Doherty, P., Furness, J., Williams, E.J. and Walsh, F.S. (1994) J. Neurochem. **62**, 2124–2131
24. Greene, L.A. and Tischler, A.S. (1982) Adv. Cell. Biol. **3**, 373
25. Yamada, K.M., Spooner, B.S. and Wessells, N.K. (1971) J. Cell Biol. **49**, 614–635
26. Letourneau, P.C. (1983) J. Cell Biol. **97**, 963–973
27. Goldberg, D.J. and Wu, D.-Y. (1995) J. Neurobiol. **27**, 553–560
28. Connolly, J.L., Green, S. and Greene, L.A. (1981) J. Cell Biol. **90**, 176–180
29. Seeley, P.J. and Greene, L.A. (1983) Proc. Natl. Acad. Sci. U.S.A. **80**, 2789–2793

30. Aletta, J.M. and Greene, L.A. (1988) J. Neurosci. **8**, 1425–1435
31. Sobue, K. and Kanda, K. (1989) Neuron **3**, 311–319
32. Kaplan, D.R., Martin-Zanca, D. and Parada, L.F. (1991) Nature (London) **350**, 158–160
33. Ohmichi, M., Decker, S.J. and Saltiel, A.R. (1992) J. Biol. Chem. **267**, 21601–21606
34. Tsao, H., Aletta, J.M. and Greene, L.A. (1990) J. Biol. Chem. **265**, 15471–15480
35. Schanen-King, C., Nel, A., Williams, L.K. and Landreth, G. (1991) Neuron **6**, 915–922
36. Hammarback, J.A. and Letourneau, P.C. (1986) Dev. Biol. **117**, 655–662
37. Kapfhammer, J.P. and Raper, J.A. (1987) J. Neurosci. **7**, 201–212
38. Bandtlow, C., Zachleder, T. and Schwab, M.E. (1990) J. Neurosci. **10**, 3837–3848
39. Caudy, M. and Bentley, D. (1986) J. Neurosci. **6**, 1781–1795
40. O'Connor, T.P., Duerr, J.S. and Bentley, D. (1990) J. Neurosci. **10**, 3935–3946
41. Sabry, J.H., O'Connor, T.P., Evans, L. et al. (1991) J. Cell Biol. **115**, 381–395
42. Bentley, D. and Toroian-Raymond, A. (1986) Nature (London) **323**, 712–715
43. Wu, D.-Y. and Goldberg, D.J. (1993) J. Cell Biol. **123**, 653–664
44. Wu, D.-Y., Wang, L.-C., Mason, C.A. and Goldberg, D.J. (1996) J. Neurosci., in the press
45. Mitchison, T. and Kirschner, M. (1988) Neuron **1**, 761–772
46. Forscher, P. and Smith, S.J. (1988) J. Cell Biol. **107**, 1505–1516
47. Okabe, S. and Hirokawa, N. (1991) J. Neurosci. **11**, 1918–1929
48. Rivas, R.J., Burmeister, D.W. and Goldberg, D.J. (1992) Neuron **8**, 107–115
49. Schweiger, A., Mihalache, O., Ecke, M. and Gerisch, G. (1992) J. Cell Sci. **102**, 601–609
50. Tsukita, S., Hieda, Y. and Tsukita, S. (1989) J. Cell Biol. **108**, 2369–2382
51. Sato, N., Funayama, N., Nagafuchi, A., et al. (1992) J. Cell Sci. **103**, 131–143
52. Schlessinger, J. and Ullrich, A. (1992) Neuron **9**, 383–391
53. Koch, C.A., Anderson, D., Moran, M.F., Ellis, C. and Pawson, T. (1991) Science **252**, 668–674
54. Reszka, A.A., Hayashi, Y. and Horwitz, A.F. (1992) J. Cell Biol. **117**, 1321–1330
55. Johansson, M.W., Larsson, E., Luning, B., Pasquale, E.B. and Rouslahti, E. (1994) J. Cell Biol. **126**, 1299–1309
56. Schmidt, C.E., Horwitz, A.F., Lauffenburger, D.A. and Sheetz, M.P. (1993) J. Cell Biol. **123**, 977–991
57. Tsui, H.-C.T., Lankford, K.L. and Klein, W.L. (1985) Proc. Natl. Acad. Sci. U.S.A. **82**, 8256–8260
58. Schaller, M.D., Hildebrand, J.D., Shannon, J.D. et al. (1994) Mol. Cell. Biol. **14**, 1680–1688
59. Schaller, M.D. and Parsons, J.T. (1994) Curr. Opin. Cell Biol. **6**, 705–710
60. Tsukita, S., Oishi, K., Akiyama, T. et al. (1991) J. Cell Biol. **113**, 867–879

3

The role of protein tyrosine kinases in transducing signals from cell adhesion molecules to promote neurite outgrowth

T.J. Baldwin, F.S. Walsh and P. Doherty*

Department of Experimental Pathology, UMDS Guy's Hospital, London Bridge, London SE1 9RT, U.K.

Introduction

The functioning of the nervous system depends on the appropriate actions of distinct neural circuits [1]. These neural circuits function because the neurons comprising them are connected appropriately to each other. One of the central themes in neurobiology is understanding the molecular mechanisms that regulate the establishment of this intricate pattern of appropriate neuronal connections during development because of its fundamental importance to the operation of the nervous system. The establishment of this pattern of neuronal connections is dependent on a number of events during development of the nervous system. One such key event is the guidance of the axon to its target via the action and orientation of the growth cone. We are interested in the molecular mechanisms that regulate axon outgrowth as this contributes significantly to establishing the specificity of neuronal connections. Initial outgrowth of axons early in development is stereospecific and depends on receptor molecules in the growth cone recognizing and transducing extrinsic cues in their environment that define a pathway [1]. Axonal pathfinding probably requires a number of different cues stimulating various signal transduction pathways that can stimulate or inhibit axonal outgrowth [2–8]. These pathway-determining cues have been demonstrated to include neurotrophic and survival factors secreted by intermediate and final targets, components of the extracellular matrix e.g. laminin, and cell adhesion molecules (CAMs) of the immunoglobulin superfamily and cadherin family [5]. In our studies we have focused on elucidating the molecular mechanisms underlying contact-dependent axonal growth on cellular substrates promoted by members of the immunoglobulin superfamily of CAMs, mainly neural cell adhesion molecule

*To whom correspondence should be addressed.

(NCAM) [9] and L1 [5], and N-cadherin [10] of the family of calcium-dependent adhesion molecules.

CAMs promoting neurite outgrowth

The observations that different cellular substrates exhibit distinct differences in their ability to support neuronal survival and axon outgrowth suggested that sets of cell-surface molecules could regulate aspects of axonal pathfinding [11,12]. The isolation and characterization of nervous system expression of CAMs of the immunoglobulin subfamily, e.g. NCAM, and of the calcium-dependent cadherin family, e.g. N-cadherin, suggested that these molecules may contribute to the regulation of axonal outgrowth during development. In general these proteins function via trans-homophilic interactions between substrate cells and neurons [6,7]. Direct evidence that these CAMs can act as a neurite outgrowth-promoting ligand in a cellular substrata comes from studies where neurons were cultured on monolayers of control and transfected fibroblasts which express physiological levels of these CAMs [13–20]. In addition indirect evidence implicating these molecules in regulating neural development comes from *in vivo* studies where the functions of the molecules have been perturbed by injection of a blocking antibody into developing embryos, resulting in abnormal neuronal development [5]. As these molecules have been characterized biochemically and functionally as adhesive molecules, these original studies tended to interpret the action of these molecules as mediating differential adhesion of the growth cone to the substrate in order to provide the primary mechanism promoting axonal outgrowth. Although, this remains a possible mode of action of these molecules in the development of the nervous system, subsequent studies implicated functions for these molecules other than simple cell adhesion. It was suggested from studies from neuronal cell cultures [16,21] and from *Drosophila* genetics [22] that these CAMs could promote development of axonal pathways by the activation of intracellular signal transduction cascades and second messengers.

Basis of a second messenger hypothesis of neurite outgrowth promoted by CAMs

Evidence from a series of experiments, using bioassays of cultured neurons on CAM-transfected monolayers, supports the notion that CAMs may promote axonal development via activation of intracellular second messengers. We have found that removal of polysialic acid (PSA), which has been reported to increase adhesion [23], from NCAM substantially inhibits NCAM-dependent neurite outgrowth [24]. These results cannot be explained simply by an inverse relationship of neurite outgrowth to adhesiveness, because we have observed positive cooperative and saturable dose–response curves for NCAM level and neurite outgrowth [18,24]. Furthermore these results are not obviously consistent with PSA operating by inhibiting the trans-binding of NCAM, as this would be expected to inhibit NCAM-dependent neurite outgrowth. These results do suggest that N-CAM-dependent neurite outgrowth and adhesion can operate separately. This can be explained by NCAM-dependent activation of a second messenger pathway rather than adhesion itself underlying neurite outgrowth response, i.e. there is no direct correlation between the ability of N-CAM to support adhesion and its ability to support neurite outgrowth. Therefore target recognition and signal transduction may be more important than adhesion in promoting axonal development stimulated by CAMs.

To test this notion a variety of agents were screened for their ability to block selectively CAM-dependent neurite outgrowth [16]. In summary, antagonists of L- and N-type calcium channels can inhibit CAM (NCAM, N-cadherin and L1)-dependent neurite outgrowth in primary neurons and pheochromocytoma cells (PC12 cells) [16,19]. This implies a role for extracellular calcium in these processes and suggests that a signal transduction cascade operates to open calcium channels. Further evidence for the involvement of calcium channels arises from the observations that potassium depolarization of PC12 cells can fully mimic CAM-dependent neurite outgrowth [25], which can be inhibited by N- and L-type calcium-channel antagonists. In addition a similar response could also be induced by the calcium-channel agonist BAY K 8644 [25]. The major conclusion from these original studies implicated an essential role for calcium influx via activation of voltage-sensitive channels into the growth cone as a result of the binding of CAMs. Such observations are consistent with the evidence implicating changes in the levels of intracellular calcium in regulating growth cone navigation [26,27]. In these experiments evidence implicated the involvement of a protein kinase in early signal transduction events and the possible activation of calcium channels via a G-protein [16].

Evidence for involvement of protein tyrosine kinases (PTKs) in neurite outgrowth promoted by CAMs

A number of observations have demonstrated that receptor and non-receptor tyrosine kinases appear to play a vital role in axonal growth and guidance. Soluble trophic factors such as nerve growth factor, brain-derived neurotropic factor (BDNF), fibroblast growth factor (FGF) and other neurotrophins function by binding and activating receptor PTKs [28]. The importance of such receptor PTKs in neural development is illustrated by the phenotypes observed after creating null mutants of the genes encoding these tyrosine kinases in mice [29]. In addition transfection of neurons with constitutively active and inactivating (dominant negative) forms of non-receptor PTKs have implicated these molecules in activating neurite outgrowth [30]. Nevertheless, null mutants of the non-receptor PTK genes in mice created by homologous recombination do not reveal clear phenotypic abnormalities associated with axonal development [31], as with the mutants in receptor PTKs. Pharmacological manipulation of neuronal cultures with inhibitors of protein kinases [16] and studies in *Drosophila* genetics [22] first identified a link between the binding of CAMs to neurons with the requirement of PTK activity in the responding neuron. This was intriguing since these CAMs do not have defined kinase domains in their sequences and thus implicated an interaction with an effector molecule. Further investigation of the mechanisms mediating axonal outgrowth on fibroblasts expressing physiological levels of CAMs involved screening inhibitors of PTKs for their ability to selectively block neurite outgrowth promoted by CAMs expressed on the surface of fibroblasts [32]. These studies demonstrated that neurite outgrowth stimulated by CAMs requires the activity of a PTK sensitive to inhibition by an analogue of erbstatin. In contrast inhibitors such as lavendustin A and genistein failed to produce inhibition of neurite outgrowth promoted by CAMs at concentrations demonstrated to inhibit the activity of non-receptor PTKs [32,33].

A clue to the potential nature of this tyrosine kinase came from experiments of Saffell et al. [34]. In these experiments an isoform of NCAM containing the VASE (variable alternative spliced exon) sequence [18,20] was expressed in PC12 cells grown on fibroblasts expressing the non-VASE isoform. In these experiments the PC12 cells lost their ability to respond to NCAM in the cellular substrate. From previous studies it had been demonstrated that NCAM VASE expressed as a substrate for neurite outgrowth was less supportive of outgrowth than non-VASE NCAM [18]. A surprising observation was that synthetic peptides containing the VASE sequence also inhibited NCAM responses at a time when NCAM itself did not contain the VASE sequence. In screening a database of protein sequences it was discovered that the VASE sequence contained 10 amino acids in common with the FGF receptor (FGFR) family of receptor PTKs [34,35] (Fig. 1). In addition, adjacent to the VASE homology sequence in the FGF receptor was the motif HAV [36,37]. The HAV sequence is an amino acid motif contained within the extracellular domain of many cadherins and is thought to play a role in the homophilic binding of cadherins expressed in different cells [36]. Nevertheless it appears that the amino acids adjacent to the HAV sequence determine the specificity of binding between cadherins. During further searches for sequence similarity a sequence between the third and fourth immunoglobulin-like domains of the CAM L1 was detected in the FGFR1 sequence containing the homologies to NCAM and N-cadherin (Fig. 1) [35]. This region in the FGF receptor has been dubbed the CAM homology domain (CHD) [35]. This sequence lies adjacent to the acid box common to FGF receptors and is well conserved between FGFR1 and FGFR2. The identification of this sequence led to the speculation that this may represent a site of interaction between neuronal FGF receptors and CAMs which could lead to the activation of this tyrosine kinase.

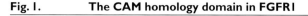

Fig. I.   **The CAM homology domain in FGFRI**

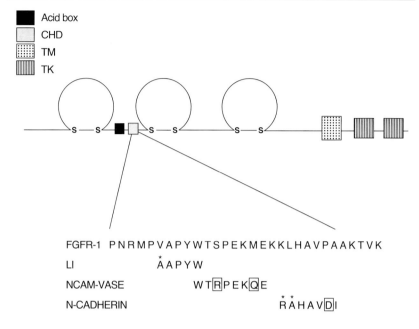

Schematic representation of FGFR1 illustrating the CHD with regions of amino acid sequence similarity identified in NCAM, L1 and N-cadherin [35].The asterisk denotes conservative amino acid changes whereas boxes denote non-conservative changes.

This postulate was examined in several ways [35]. First, an antiserum to the acid box of the FGF receptor was tested for its ability to disrupt neurite outgrowth supported by NCAM, L1 and N-cadherin expressed in 3T3 cells. This antiserum demonstrated no detectable effect on basal, integrin-mediated, neurite outgrowth on parental 3T3 monolayers. In contrast, on 3T3 cells expressing CAMs the anti-(FGF receptor acid box) antibody completely inhibited the response to L1 and partially inhibited (~ 60%) the response to NCAM and N-cadherin. Secondly, an antiserum raised against the CHD peptide sequence from the FGF receptor inhibited fully and selectively the neurite outgrowth response stimulated by NCAM, L1 and N-cadherin. Furthermore, pretreatment of the neurons with the anti-CHD antibody prior to plating on the 3T3 monolayers expressing CAMs suggests that the inhibition observed resulted from inactivation of the FGF receptor in the responding neuron. Thirdly, peptides derived from the FGFR1 CHD selectively inhibit neurite outgrowth responses stimulated by CAMs. Addition of a 31-mer peptide containing the CHD blocked neurite outgrowth responses to NCAM, L1 and N-cadherin in a dose-dependent manner. The basal outgrowth on

3T3 cells was unaffected by addition of this peptide and addition of a scrambled version of this sequence to cultures where 3T3 cells expressed CAMs had no effect on CAM-stimulated neurite outgrowth. In addition, peptides corresponding to overlapping subdomains of this sequence selectively blocked CAM-mediated neurite outgrowth. In this case, however, the peptides selectively inhibited the response corresponding to the CAM with which they exhibit most identity (Fig. 1). Similarly peptides derived from the equivalent domains of FGFR2, which is also expressed in the nervous system, demonstrated equivalent responses in terms of their effectiveness and specificity for blockade of neurite outgrowth responses to CAMs.

In addition FGF was demonstrated to stimulate neurite outgrowth from cerebellar granule cells by utilizing a second messenger pathway that was pharmacologically indistinguishable from that activated by CAMs expressed in 3T3 cells [38] (see below). As with the CAMs, antibodies to the acid box sequence and CHD, as well as peptides from the CHD can inhibit neurite outgrowth stimulated by FGF [38]. In addition a peptide derived from basic FGF (KRTGQYKL), which has been shown previously to block FGF binding to its receptor, blocked neurite outgrowth promoted by FGF but had no effect on neurite outgrowth stimulated by CAMs [35]. This suggests that there is some specificity in the apparent response of the FGF receptor to the CAMs and that neurite outgrowth stimulated by CAMs does not require addition of FGF. The effect of all of these blockade experiments could be alleviated by use of mimetics that activate downstream components of the pathway activated by CAMs, e.g. depolarization by K^+, which suggests that the FGF receptor is important as a very early component of the signal transduction pathway used for CAMs in promoting neurite outgrowth.

Evidence to support the activation of FGF receptors by binding of CAMs is provided by the use of recombinant CAM–Fc chimeras [35]. These molecules represent fusions of the extracellular domain of CAMs to the Fc portion of human immunoglobulin [39]. When such fusions are transfected into fibroblastic cells large amounts of soluble dimeric proteins are secreted into the tissue-culture supernatants which can then be affinity-purified by binding to and elution from columns bearing Protein-A. Addition of FGF (25 ng/ml) or L1–Fc chimera (1 μg/ml) for 1 h to cultures of cerebellar granule cells plated on polylysine resulted in the increase in tyrosine phosphorylation of a number of common cellular proteins when assayed by Western blotting using anti-phosphotyrosine antibodies [35]. This increase in apparent tyrosine phosphorylation of a common set of proteins, as a result of FGF or L1 treatment, could be inhibited by pretreatment of neurons with the antibodies to the CHD. Unlike previous data which suggested a requirement for tyrosine kinase activity for CAM-stimulated neurite outgrowth these results are the first to provide evidence for actual activation of tyrosine kinases in neurons in response to addition of a CAM. Although no direct characterization of the proteins phosphorylated was achieved these data are suggestive of activation of the FGF receptor by L1 because of the commonality of proteins phosphorylated.

The results of subsequent experiments suggest that the soluble L1–Fc chimera could stimulate neurite outgrowth in cerebellar granule cells as effectively as FGF or L1 expressed in 3T3 cells [40]. The activity of the L1–Fc chimera was maximal at addition of 5 µg/ml to cultures and could be blocked by pretreatment of neurons with antibodies to L1 or antibodies to the FGF receptor. Furthermore, use of pharmacological tools that block components of the signal transduction pathway downstream of the FGF receptor (see below) also block neurite outgrowth promoted by soluble L1 [40].

In summary the above results suggest that NCAM, L1 and N-cadherin may stimulate neurite outgrowth in neurons by activating FGF receptor tyrosine kinases. These results also establish that FGF receptors contain a possible recognition sequence, the CHD, which is important for the function of L1, NCAM and N-cadherin. One model suggested by these results for the mechanism of activation is illustrated in Fig. 2. In this model the trans-homophilic binding of the CAM would lead to recruitment of CAMs at a point of cell–cell contact. This clustering of CAMs could lead to a co-clustering of an effector molecule such as the FGF receptor. This could be aided by a direct or indirect binding of the CAM to the FGF receptor (*cis/trans* or both). In such models the peptides could function to block sites present on the CAMs rather than the FGF receptor. Nevertheless, the fact that these peptides also block FGF activation of the FGF receptor suggest that they do not merely function to block homophilic binding.

Relationship of tyrosine kinases to activation of downstream pathway events involved in transducing signals from CAMs

The experiments outlined above demonstrate two components of the signal transduction pathway characterized in neurons as a response to stimulation by binding CAMs. Initial steps require FGF receptor activity and stimulation culminates in calcium influx into neurons [6]. A similar influx of calcium via voltage-dependent calcium channels appears to be a requirement for neurite outgrowth promoted by FGF in cerebellar granule cells [38]. A clue to the pathway of activation of these molecules has been derived from the use of the diacylglycerol (DAG) lipase inhibitor RHC-80267 [38]. Addition of the DAG lipase inhibitor to neuronal cultures specifically inhibits neurite outgrowth stimulated by CAMs expressed in 3T3 cells, FGF and L1–Fc chimeras [38,40,41]. DAG lipase catalyses the conversion of DAGs into fatty acids such as arachidonic and oleic acid [38]. Furthermore, arachidonic and oleic acids have been identified as agents that can activate the opening of voltage-dependent, calcium channels [42]. Thus, mobilization of these fatty acids in the neuron could provide a means for activation of calcium channels in response to CAM stimulation. In support of this hypothesis application of arachidonic or oleic acid to neuronal cultures can produce neurite

Fig. 2.   **A putative model for CAM/FGF receptor interactions**

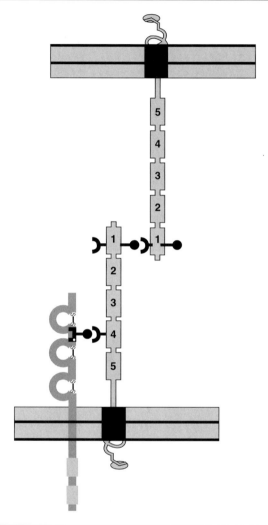

Schematic diagram illustrating homophilic trans-binding of two CAMs on different cells leading to the cis-activation of an FGF receptor in the responding neuron.

outgrowth characteristic of that stimulated by CAMs [41]. This observation is supported by experiments where addition of mellitin, which mobilizes arachidonic acid in cells, promotes neurite outgrowth [41]. An interesting observation was made in the study of the FGF stimulation of neurite outgrowth in cerebellar granule cells. A biphasic dose–response curve of neurite outgrowth to FGF application was observed (Fig. 3a). Neurite outgrowth was stimulated maximally by application of 5 ng/ml FGF to the culture, the response declining at higher doses [38]. This biphasic response, however, did not appear to result from a classical desensitization of the FGF receptor, as tyrosine phosphorylation of cellular proteins was apparent at all doses of FGF [38]. A possible resolution of this apparent conundrum is provided by the observation that a biphasic neurite outgrowth response curve was observed on application of arachidonic acid [38,41]. The maximal outgrowth was observed at 10 μM arachidonic acid and a decline in responsiveness was observed at higher doses [38,41] (Fig. 3b). This suggests that the signal transduction mechanism downstream of the FGF receptor becomes 'saturated' or desensitized in response to CAM stimulation, thereby restricting responsiveness in terms of neurite outgrowth.

The involvement of DAG lipase in the signal transduction pathway implicates phospholipase C γ (PLCγ) as an earlier event in the signal transduction pathway, as arachidonic acid containing DAG is one product of the activated PLCγ hydrolysis of phospholipids with the other product of PLCγ activity being inositol phosphates. One potential signal for the activation of PLCγ is the activation of FGF receptor, which when activated becomes autophosphorylated and can bind PLCγ via its SH2 domain and subsequently activate it by phosphorylation [43]. A summary of these observations is illustrated in the pathway outlined in Fig. 4. Thus the observations relating to identification of second messenger components mediating CAM binding to calcium mobilization may fit the model for FGF receptor activation [35].

A key event in CAM-stimulated neurite outgrowth is the influx of calcium into the neuron via voltage-activated calcium channels. We have identified a calcium calmodulin-dependent protein kinase (CamK) as a key element in transducing this calcium signal to axonal growth [44]. Neurite outgrowth promoted by NCAM, L1, N-cadherin, FGF, arachidonic acid and K^+ depolarization is specifically inhibited by dose-dependent application of the CamK inhibitor KN-62 to neuronal cultures [45]. CamK has been implicated in regulating aspects of synaptic plasticity [44], regulating the organization of the neuronal cytoskeleton [44] and, on transport into the nucleus, activating transcription of genes such as *cfos* [44]. Thus the postulated signal transduction pathway illustrated in Fig. 4 can lead to the activation of an enzyme that has been implicated in activating a number of intracellular events associated with axonal outgrowth.

Future prospects

As a result of our observations a new set of questions arise concerning the mechanisms of signal transduction activated by CAM-stimulated neurite outgrowth. We have proposed a rather radical postulate that CAMs such as NCAM, L1 and N-cadherin can stimulate neurite outgrowth by *cis*-activation of an FGF receptor PTK in a neuron in response to trans-homophilic binding or binding

Fig. 3.   Dose–response for FGF and arachidonic acid activation of neurite outgrowth

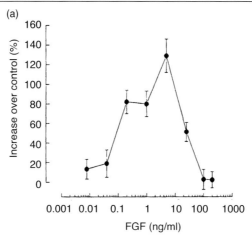

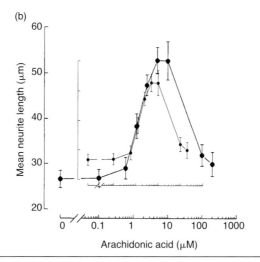

(*a*) FGF stimulates neurite outgrowth from cerebellar granule cells. Cerebellar neurons plated on 3T3 cells and treated with a range of FGF concentrations. The mean neurite length was determined and plotted as a percentage increase in neurite length observed when plated on 3T3 cells. (*b*) Arachidonic acid (AA) stimulates neurite outgrowth from cerebellar neurons in a dose-dependent manner. Cerebellar granule cells were cultured on 3T3 cells and neurites measured as in (*a*) but treated with a range of concentrations of AA.

of an appropriate ligand on a substrate cell. This represents activation of a common signalling mechanism by quite distinct ligands. In contrast, the experiments of

Fig. 4. **Model of signal transduction pathway used by CAMs via FGF receptor activation to promote neurite outgrowth**

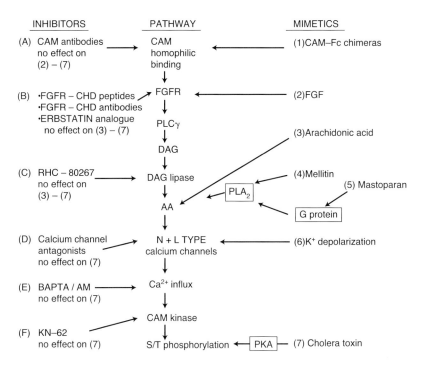

A linear representation of the steps involved in the CAM second messenger pathway. The central column lists the steps in the pathway determined by inhibitors and mimetics. (A)–(F) denote specific inhibitors, (1)–(7) denote the reagents that can directly activate components of the pathway.

Maness and collaborators [46,47] suggest that CAMs such as L1 and NCAM activate neurite outgrowth by activation of non-receptor PTKs. In this case each CAM has a corresponding specific tyrosine kinase, e.g. NCAM is specific for fyn and L1 is specific for src. In these experiments cerebellar granule cell neurons from mice, in which the tyrosine kinases had been inactivated by use of homologous recombination, were CAM-expressing substrates. Although, some of the experimental paradigms used in measuring stimulation of neurite outgrowth are distinct from ours it represents an intriguing finding. First, what is the proposed mechanism of activation of these non-receptor PTKs by CAMs? Although there is some evidence from co-immunoprecipitation assays, indicating that glycosylphosphatidylinositol-linked CAMs such as Thy-1 interact with non-receptor PTKs in T-cells [48], this does not explain the mechanism of activation of the tyrosine kinase. In experiments of this nature we are still measuring a requirement for the expression of these proteins. A loss of expression of these tyrosine kinases may result in the loss of expression or alteration of activity of other molecules that may be required for CAM-mediated neurite outgrowth. In itself it does not represent evidence for activation of these tyrosine kinases by CAM. Furthermore, *in vivo* 'knockouts' of these tyrosine kinases do not result in major alterations of the development of axonal projections [31]. In the case of fyn⁻ mice there are alterations in the development of the hippocampus and defects in LTP [31]. It is of interest that null mutants of NCAM also demonstrate defects in spatial learning [49,50]. In light of the latter observations it is possible that fyn may have a role in mediating aspects of plasticity requiring CAM-mediated axonal growth; nevertheless, mutations in src do not exhibit a similar phenotype despite its apparent requirement for substrate-bound L1-stimulated neurite outgrowth. These experiments involving 'gene specific' inhibitions demonstrate a clear approach to testing the postulate that FGF receptors are required to mediate CAM-stimulated axon outgrowth. In contrast, to non-receptor tyrosine kinases germ-line null mutations in FGFR1 result in early embryonic lethality [51,52]; therefore alternative approaches must be developed to examine its role in axonal development. The clearest approach is to use the strategy of expressing 'dominant negative' mutants of the FGF receptor [53] in cultured neurons and also specifically in the nervous system of transgenic mice to specifically inactivate the FGF receptor function. This approach will test the postulate that the FGF receptor is required for CAM-mediated outgrowth in cell culture and that it is important as a molecular mechanism for determining axonal outgrowth during development.

Further tests of the postulate involve demonstrating that CAMs as well as FGF can bind directly to the FGF receptor. Such an analysis must include a number of CAMs, in particular N-cadherin, since other effector molecules, the catenins, have been identified as transducers of extracellular signals to the cellular cytoskeleton [54]. Also, as a consequence of this the FGF receptor itself becomes phosphorylated and enzymes such as PLCγ become phosphorylated and activated to produce second messenger molecules such as arachidonic acid and inositol phosphates. It is clear that the production of pure soluble chimeric CAMs that can mimic stimulation of neurite outgrowth promoted by CAMs expressed in 3T3 cells are invaluable tools for making these experiments feasible. To date our evidence implicates the activation of PLCγ by the FGF receptor as sufficient and necessary to transduce signals promoting neurite outgrowth as a result of binding CAMs. This observation raises intriguing questions concerning the potential nature of signalling from the FGF receptor by CAMs and FGF in neurons. First, is the mechanism for activation of the FGF receptor the same in each case? Is there a requirement for dimerization or can the tyrosine kinase domain be activated via another mechanism? Secondly, there is a great deal of evidence suggesting that p21^{ras} and mitogen-activated protein kinases are activated by receptor tyrosine kinases, including FGF receptors, in neurons to promote neuronal differentiation [55]. It will be of great interest to determine whether such a pathway is activated by, or required for, CAM-mediated axonal outgrowth. It is possible that this may vary between different neuronal types and mechanisms of activation of the FGF receptor.

Our observations suggest that activation of FGF receptors can occur as a result of binding of CAMs to neurons. This is not only of interest to those involved in understanding the molecular mechanism underlying axonal development mediated by CAMs but also has important implications in interpreting the actions of FGF during development. It is possible that certain events in development attributed to FGFs may be mediated by CAMs activating FGF receptors.

This work was supported by grants from the MRC UK, the Wellcome Trust, and the Dunhill Trust. P.D. is supported by an MRC Senior Fellowship. We thank H. Rickard for help with the manuscript.

References

1. Purves, D. and Lichtman, J.W. (1985) Principles of Neural Development. Sinauer Associates, Sunderland, MA, U.S.A.
2. Dodd, J. and Jessell, T.M. (1988) Science **242**, 692–699
3. Doherty, P. and Walsh, F.S. (1989) Curr. Opin. Cell Biol. **1**, 1102–1106
4. Lumsden, A. and Cohen, J. (1991) Curr. Opin. Genet. Dev. **1**, 230–235
5. Bixby, J.L. and Harris, W.A. (1991) Annu. Rev. Cell Biol. **7**, 117–159
6. Doherty, P. and Walsh, F.S. (1992) Curr. Opin. Neurobiol. **2**, 595–601
7. Rutishauser, U. (1993) Curr. Opin. Neurobiol. **3**, 709–715
8. Dodd, J. and Schuchardt, A. (1995) Cell **81**, 471–474
9. Edelman, G.M. (1985) Annu. Rev. Biochem. **54**, 135–169
10. Takeichi, M. (1991) Science **251**, 1451–1455
11. Bixby, J.L., Pratt, R.S., Lillien, J. and Reichardt, L.F. (1987) Proc. Natl. Acad. Sci. U.S.A. **84**, 2555–2559
12. Neugebauer, J.K.M., Tomaselli, K.J., Lillien, J. and Reichardt, L.F. (1988) J. Cell Biol. **107**, 1177–1187
13. Matsunaga, M., Hatta, N., Nagafuchi, A. and Takeichi, M. (1988) Nature (London) **334**, 62–64
14. Doherty, P., Barton, C.H., Dickson, G. et al. (1989) J. Cell Biol. **109**, 789–798
15. Doherty, P., Fruns, M., Seaton, P. et al. (1990) Nature (London) **343**, 464–466
16. Doherty, P., Ashton, S.V., Moore, S.E. and Walsh, F.S. (1991) Cell **67**, 21–33
17. Doherty, P., Rowett, L.H., Moore, S.E., Mann, D.A. and Walsh, F.S. (1991) Neuron **6**, 247–258
18. Doherty, P., Moolenaar, C.E.C.K., Ashton, S.V., Michalides, R.J.A.M. and Walsh, F.S. (1992) Nature (London) **356**, 791–793
19. Williams, E.J., Doherty, P., Turner, G. et al. (1992) J. Cell Biol. **119**, 883–892
20. Liu, L., Haines, S., Shaw, R. and Akeson, R.A. (1993) J. Neurosci. Res. **35**, 327–345
21. Schuch, E., Lohse, M.J. and Schachner, M. (1989) Neuron **3**, 13–20
22. Elkins, T., Zinn, K., McAllister, L., Hoffman, F.M. and Goodman, C.S. (1990) Cell **60**, 565–575
23. Hoffman, S. and Edelman, G.M. (1983) Proc. Natl. Acad. Sci. U.S.A. **80**, 5762–5766
24. Doherty, P., Cohen, J. and Walsh, F.S. (1990) Neuron **5**, 209–219
25. Saffell, J.L., Walsh, F.S. and Doherty, P. (1992) J. Cell Biol. **118**, 663–670
26. Kater, S.B. and Mills, L.R. (1991) J. Neurosci. **11**, 891–899
27. Silver, R.A., Lamb, A.G. and Bolsover, S.R. (1990) Nature (London) **343**, 751–754
28. Chao, M.V. (1992) Neuron **9**, 583–593
29. Barbacid, M. (1994) J. Neurobiol. **25**, 1386–1403
30. Maness, P.F. (1992) Dev. Neurosci. **14**, 257–270
31. Grant, S.G., O'Dell, T.J., Karl, K.A. et al. (1992) Science **258**, 1903–1910
32. Williams, E.J., Walsh, F.S. and Doherty, P. (1994) J. Cell Biol. **124**, 1029–1037
33. Bixby, J.L. and Jhabvala, P. (1992) J. Neurobiol. **23**, 468–480

34. Saffell, J.L., Walsh, F.S. and Doherty, P. (1994) J. Cell Biol. **125**, 427–436
35. Williams, E.J., Furness, J., Walsh, F.S. and Doherty, P. (1994) Neuron **13**, 583–594
36. Blaschuk, O., Sullivan, R., David, S. and Pouliaut, Y. (1990) Dev. Biol. **139**, 227–229
37. Byers, S., Amaya, E., Munro, S. and Blaschuk, O. (1992) Dev. Biol. **152**, 411–414
38. Williams, E.J., Furness, J., Walsh, F.S. and Doherty, P. (1994) Development **120**, 1685–1693
39. Simmons, D.L. (1993) in Cellular Interactions in Development. A Practical Approach (Hartley, D.A., ed.), pp. 93–127, IRL Press, Oxford, U.K.
40. Doherty, P., Williams, E.J. and Walsh, F.S. (1995) Neuron **14**, 57–66
41. Williams, E.J., Walsh, F.S. and Doherty, P. (1994) J. Neurochem. **62**, 1231–1234
42. Divecha, N. and Divine, R.F. (1995) Cell **80**, 269–278
43. Schlessinger, J. (1994) Curr. Opin. Gen. Dev. **4**, 25–30
44. Williams, E.J., Mittal, B.M., Walsh, F.S. and Doherty, P. (1995) Mol. Cell. Neurosci. **1**, 69–79
45. Ghosh, A. and Greenberg, M.E. (1995) Science **268**, 239–246
46. Beggs, H.E., Soriano, P. and Maness, P.F. (1994) J. Cell Biol. **127**, 825
47. Ignelzi, M.A., Miller, D.R., Soriano, P. and Maness, P.F. (1994) Neuron **12**, 873–884
48. Stefanova, I., Horehsi, V., Ansotegni, I.J., Knapp, W. and Stockinger, H. (1991) Science **254**, 1016–1019
49. Cremer, H., Lange, R., Christoph, A. et al. (1994) Nature (London) **367**, 455–459
50. Tomasiewicz, H., Ono, K., Yee, D. et al. (1993) Neuron **11**, 1163–1174
51. Yamaguchi, T.P., Harpal, K., Henkemayer, M. and Rossant, J. (1994) Genes Dev. **8**, 3032–3044
52. Deng, C.X., Wynshaw-Boris, A., Shen, M.M. (1994) Genes Dev. **8**, 3045–3057
53. Amaya, E., Musci, T.J. and Kirschner, M.W. (1991) Cell **66**, 257–270
54. Ranscht, B. (1994) Curr. Opin. Cell Biol. **6**, 740–746
55. Marshall, C.J. (1995) Cell **80**, 79–85

4

Microtubule function in growth cones

Maxwell S. Bush, Mandy Johnstone and Phillip R. Gordon-Weeks*

The Randall Institute, King's College London, 26–29 Drury Lane,
London WC2B 5RL, U.K.

Introduction

Neural development is crucially dependent on the ability of growth cones to navigate precise routes through the embryo, a process generally referred to as 'pathfinding', and to recognize appropriate target cells with which to form a synapse. Growth cones are sensitive to molecular guidance cues, such as chemotropic factors, cell adhesion proteins and extracellular matrix molecules, that influence the motile behaviour of the growth cone and hence the direction of axon growth. These cues interact with growth cone membrane receptors and lead, via intracellular signalling events, to changes in the growth cone cytoskeleton and hence in directional motility. In this context the growth cone cytoskeleton is the 'final common path of action' of extrinsic guidance cues. Despite considerable effort to identify and characterize extrinsic guidance molecules [1–4], compared with other motile cells we have only a rudimentary knowledge of how guidance cues regulate the growth cone cytoskeleton. In this review we will assess the role of the growth cone cytoskeleton in pathfinding with particular emphasis on microtubules and their interactions with filamentous actin (F-actin).

The growth cone cytoskeleton

The major filamentous components of the cytoskeleton of growth cones are microtubules and microfilaments, neurofilaments are usually absent (Fig. 1). Microtubules are prominent in the central domain (C-domain) of the growth cone, whereas microfilaments are concentrated in the peripheral (P), motile regions and in the cortical cytoskeleton.

---

*To whom correspondence should be addressed.

Fig. I.   **Diagram of the growth cone cytoskeleton**

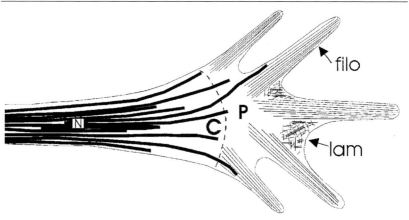

Microtubules (thick lines) are tightly bundled in the neurite (N) but splay out individually or in small bundles on entering the C-domain of the growth cone. Occassionally, individual microtubules extend across the P-domain and insert into the base of a filopodium (filo) the core of which is composed of F-actin bundles (thin lines). Between adjacent filopodia, sheet-like extensions or lamellipodia (lam) advance. Lamellipodia are supported by a meshwork of actin filaments (thin lines). The boundary between the C- and P-domains is indicated by a dotted line.

Microfilaments

As they advance, growth cones extend and retract from their motile regions finger-like processes, or filopodia, and between them, sheet-like extensions or lamellipodia. Filopodia are the first part of the growth cone to encounter new territory and as such are ideally located to carry out their principal function; the detection and initial response to guidance cues. A considerable body of evidence has accumulated to support this pathfinding role for filopodia (reviewed in [5]). For instance, filopodia contain a core-bundle of actin filaments, which are necessary for the extension and retraction of the filopodia (reviewed in [6,7]), and when filopodial extension is inhibited by the action of drugs that depolymerize F-actin, such as the cytochalasins, pathfinding is compromised [8,9]. Furthermore, filopodia have receptors for guidance cues, such as integrins, concentrated at their tips [10] and contain appropriate intracellular signalling pathways that are independent of the rest of the growth cone [11]. Precisely how filopodia relay information about the local environment to the growth cone is not clear.

Microtubules

In the axon, microtubules provide the substrate for fast axonal transport and support the cylindrical shape of the axon. The function of microtubules in growth cones is less clear. In the axon shaft, microtubules are bundled into fascicles but on entering the C-domain of the growth cone they de-fasciculate and splay out individually, or in small bundles, to cross into the P-domain of the growth cone. Observations of living growth cones in which the microtubules have been fluorescently labelled by microinjection of fluorescent tubulin, show that microtubules in the proximal regions of growth cones are continually extending into and retracting from the distal P-domain [12–14]. Microtubules in cells undergo cyclical periods of relatively slow polymerization and rapid depolymerization known as dynamic instability. This behaviour in growth cones, or polymer sliding, may explain the extension and retraction of microtubules seen in living growth cones (reviewed in [15,16]). Dynamic instability allows microtubules to continually probe and interact with the actin cytoskeleton in the P-domain of the growth cone and may underlie growth cone turning [17,18].

Interactions between microtubules and filopodia

We have observed that individual microtubules viewed by immunofluorescence in growth cones in culture occasionally extend across the P-domain and insert into the proximal part of a filopodium [19]. Recently, we have confirmed this observation at the ultrastructural level using immunogold electron microscopy [20]. We found that in growth cones of cultured dorsal root ganglion (DRG) neurons, individual microtubules traversing the P-domain occasionally have their distal ends closely apposed to the proximal part of the actin filament bundle that forms the core of the filopodium (Fig. 2). This observation suggests that there may be a specific interaction between microtubules and F-actin in growth cones.

Interactions between microtubules and F-actin in growth cones have previously been suspected on the basis of indirect evidence. For instance, when the microtubule-stabilizing drug taxol is applied to growth cones in stoichiometric concentrations (μM), microtubule polymerization and bundling is stimulated and bundled loops of microtubules form in the C-domain. Concurrently, axon elongation is blocked and the motile P-domain collapses on to the microtubule loops [21–27]. These observations suggest that microtubules contribute to the shape of the P-domain of the growth cone by an interaction with the actin cytoskeleton. Conversely, when microtubules are depolymerized, with drugs such as colchicine, ectopic growth cones appear along the axon shaft [28–30]. In some circumstances, depolymerization of the actin cytoskeleton in the growth cone can allow microtubules to extend more distally [31,32].

An interaction between microtubules and F-actin has also been inferred from direct observation of living growth cones. In the Ti1 pioneer growth cones of grasshopper limb buds growing in explant culture and microinjected with fluorescent tubulin to label microtubules in the growth cone, microtubules were seen to selectively invade or be selectively retained in filopodial branches (but not filopodia) that had developed from single filopodia by the accumulation of F-actin; these branches eventually became new neurites [12]. These authors suggested that differential organization of microtubules across the growth cone is an important component of vectorial growth. A similar conclusion was also reached by Lin and

Fig. 2.   Microtubule interactions with F-actin in growth cones

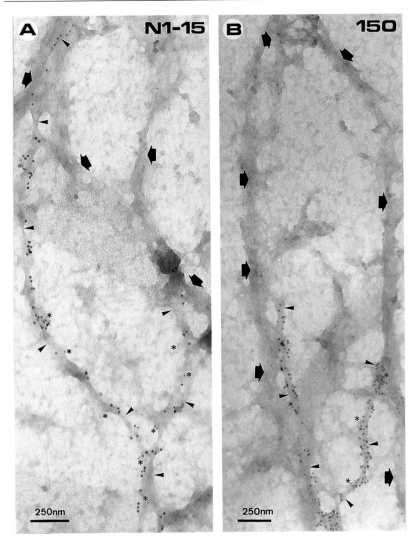

*Electron micrographs of growth cone cytoskeletons double immunogold-labelled with monoclonal antibody YL 1/2 (which recognizes α-tubulin; large gold particles) and the MAP 1B antibodies N1-15 (**A**; recognizes all forms of MAP 1B) and 150 (**B**; recognizes a mode I phosphorylation epitope on MAP 1B). The distal ends of some microtubules (arrowheads) that are labelled only at discrete points with MAP 1B antibodies (asterisks, small gold particles), lie in close association with the proximal ends of filopodial F-actin bundles (arrows).*

Forscher [32] studying *Aplysia* bag-cell growth cones interacting with each other in culture and stained by tubulin antibodies. They found that when filopodia from one growth cone contacted another growth cone, F-actin accumulated at the contact site and subsequently microtubules selectively extended into the contact region in both growth cones and became aligned along the interaction axis. This microtubule reorganization was associated with an advance of the C-domain along the direction of orientation of the microtubules, a process called engorgement. This sequence of events is consistent with the idea that stabilized filopodial F-actin can recruit microtubules to support neurite growth.

Collectively, these observations suggest that there is an interaction between microtubules and F-actin in growth cones, but what form could such an interaction take? Microtubule distribution is known to be restricted by the actin cytoskeleton. For instance, hepatoma cells transfected with microtubule-associated protein 2C accrue stiff microtubule bundles around their cell circumference [33], but when the restraining actin cytoskeleton is depolymerized with cytochalasin D, these bundles straighten out into long processes [34]. What force could overcome such restriction and reorientate microtubules to sites of accumulated actin, as seen in *Aplysia* and grasshopper Ti1 growth cones? There may be a local increase in tension generated by enhanced myosin activity in the accumulated F-actin. Experiments by Heidemann's group show that local tension applied to the cell bodies of cultured neurons can stimulate neurite growth via new microtubule polymerization [35]. Support for this model comes from video images of cultured *Aplysia* growth cones that show evidence of force (tension) transduction, such as bending of neurites and stretching of filopodia [32]. In recently plated cultured sympathetic ganglion cells that are extending their first filopodia, Smith [36] found that the movement of perinuclear cytoplasm and microtubules is directed into those filopodia that contact another cell at their tips and then develop tension; these filopodia then dilate and become neurites. Perhaps the increased tension induces a gel to sol transition in the actin cytoskeleton or microtubule advance (either by polymerization or sliding) is stimulated [35] and these factors are conducive to invasion by dynamic microtubules. Alternatively, the local accumulation of actin may leave pockets of low F-actin density, due to a contact-mediated reduction in the retrograde flow of F-actin, and this may facilitate microtubule invasion [32].

Microtubules and growth cone pathfinding

A fundamental event in growth cone turning during pathfinding is the selective stabilization of filopodia on the side of the turn (Fig. 3; reviewed in [7,37]). On the basis of our observations of the interaction of microtubules with F-actin in filopodia, and the observations discussed above, we have suggested that stabilized filopodia that have contacted guidance cues can capture and stabilize microtubules and that this event leads to a microtubule reorganization in the growth cone that is an essential step in growth cone turning (Fig. 3, [18–20]).

Fig. 3.   **Microtubule re-organization in turning growth cones**

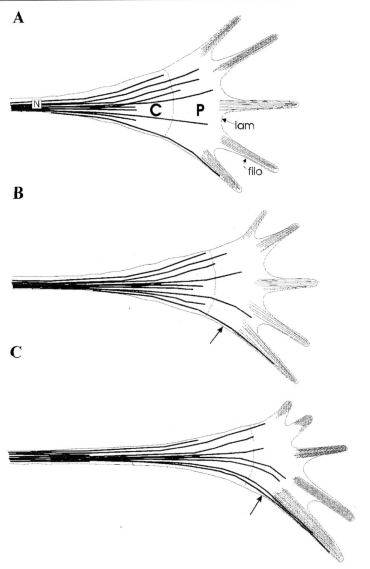

Diagrams showing the hypothetical re-organization of microtubules (thick lines) and microfilaments (thin lines) in a growth cone undergoing a turning manoeuvre. Time elapses from (A) to (C). It is imagined that the lower filopodia (filo) and lamellipodia (lam) encounter a guidance signal which stabilizes these structures. As a consequence, microtubules that are dynamically unstable and randomly probing the peripheral domain are 'captured' by filopodial F-actin and stabilized (arrows in B and C). Dashed lines indicate the boundary between the C- and P-domains.

How do only the dynamic microtubules captured by filopodia and orientated in the correct direction for turning become stabilized? Stability is most probably induced by microtubule-associated proteins (MAPs) binding to microtubules and at a later stage by post-translational modifications of tubulin. How do MAPs only recognize reorientated microtubules? The distal end of a microtubule may be transiently stabilized by interacting with the proximal end of a filopodial actin bundle, either through a MAP–actin-binding protein interaction or electrostatic attraction. The result would be to increase that microtubule's half-life, thereby enhancing the probability that soluble MAPs would bind and stabilize it. There are large, soluble pools of MAPs in the growth cone [20,26,38,39] and subsets of these MAPs bind to microtubules [20,26,40]. A gel to sol transition of the actin cytoskeleton may also prolong microtubule life by reducing the local shear and stress forces acting on the distal end of the lattice.

Another vital phase of normal growth cone pathfinding is the bundling of microtubules in the proximal growth cone and its consolidation into a new axon [12–14,17]. Bundling is seen microscopically as a lateral 'zippering' of microtubules towards the new axis of growth [13,32] and occurs prior to the constriction of the growth cone into an axon [13,14]. This suggests that the microtubules are not squeezed together by the cortical actin cytoskeleton in the proximal growth cone, but that bundling may be induced by other factors. It is not clear whether MAPs can bundle microtubules directly by cross-linking them or indirectly by stabilizing them or neutralizing their repulsive net anionic charge ([41]; reviewed in [42]). The projection domains of MAPs probably keep microtubules apart, so maintaining an inter-microtubule space for the fast axonal transport of membranous organelles [43–45]. This transport may be involved in the engorgement stage of axonogenesis, in which the cytoplasm from the C-domain surges forward into the P-domain [46].

Engorgement may be vectorially directed within the growth cone to bring about growth cone turning. Cross-linking microtubules to actin bundles raises the possibility that organelles may be preferentially transported into the filopodia or lamellipodia that are directing the turning behaviour of the growth cone [19]. Evidence for this comes from observations of individual vesicles, in extruded squid axoplasm, that can be transported along both microtubule and actin tracks, suggesting that these vesicles carry motor proteins that can use either a tubulin or an actin substrate [47].

To test the role of microtubules in growth cone turning we have exploited a recently established *in vitro* assay in which growth cones of DRG explants make right-angled turns when confronting a sharp substrate border of a non-permissive molecule [48,49]. In growth cones growing on a permissive substrate, such as laminin, microtubules are symmetrically distributed within the growth cone such that the distal ends of microtubules are approximately equally distributed on either side of an imaginary line that bisects the growth cone into two halves of equal area. In contrast, microtubules become reorganized at the substrate border such that more microtubule ends are found in one half of the growth cone [18]. This finding suggests that microtubule reorganization is associated with growth cone turning. To test this idea further we have assessed the effects of restricting the distribution of microtubules in growth cones to the C-domain and reducing their dynamic behaviour using substoichiometric (low nanomolar) concentrations of taxol. The striking finding is that taxol reversibly blocks growth cone turning at the substrate border. These results provide strong evidence for the hypothesis that microtubule reorganization in the growth cone is a necessary event before growth cones can turn under the influence of a guidance molecule.

A corollary of the hypothesis that stabilized filopodia capture microtubules is that proteins that cross-link microtubules to F-actin may be localized to the interface between microtubules and the proximal region of the filopodial actin-filament bundle. It would not be surprising if such an interaction were mediated by MAPs or actin-filament-binding proteins. There are two MAPs that are both present on growth cone microtubules and bind actin *in vitro*: tau and MAP 1B. These MAPs have both been proposed to fulfil this cross-linking function [16,40].

Tau

Extensive investigation suggests that tau is important in the assembly and stability of microtubules and that it plays a role in axonal growth (reviewed in [50–52]). However, in homozygous tau 'knockout' mice [53], the only phenotype is a difference in the numbers of microtubules and microtubule cross-bridges in parallel-fibre axons of the cerebellum and a slight up-regulation of MAP 1A. The nervous system of these mice is otherwise grossly normal and, importantly, cultured neurons extend axons at a normal rate. Such evidence suggests that tau may be functionally redundant and that its absence can be compensated for by increased levels of MAP 1A.

Tau is essentially a neuron-specific phosphoprotein concentrated in axons, but is also present in dendrites; it is heterogeneous in structure and phosphorylation status and both of these factors are major functional determinants. The molecular heterogeneity of tau is due to developmental regulation by alternative splicing of a 6 kb transcript [54] of a single gene and also by post-translational phosphorylation of several low-molecular-mass isoforms [50,55]. Addition of two sequences at the N-terminus [56] and the insertion of a fourth microtubule-binding repeat in the highly conserved C-terminus [57], generates six possible length-repeat combinations [58]. In embryonic rat brain, a single 48 kDa juvenile tau (tau_J) protein with three C-terminal repeats and no N-terminal insertions is replaced, from post-natal day 15, by six adult (tau_A) isoforms (55–68 kDa) [59]. A high-molecular-mass (HMM) form of tau (110–120 kDa), that is preferentially expressed in the adult peripheral nervous system (PNS) [60,61] has four C-terminal repeats, extra inserts in the N-terminus and is encoded by an 8 kb mRNA [62,63].

Microtubule binding and bundling

Tau stimulates the assembly of microtubules by lowering the critical concentration for tubulin, and enhancing microtubule nucleation, growth and stability. All of these effects are dependent on the ability of tau to bind electrostatically to tubulin dimers in the microtubule lattice. Synthetic tau peptides containing a single C-terminal repeat bind to microtubules [64], while four-repeat constructs assemble microtubules more actively than three-repeat forms [58], suggesting that the repeats are responsible for binding. Curiously, constructs consisting of the repeats alone bind poorly, but their binding is greatly enhanced if the flanking regions (which themselves have insignificant binding properties) are present [51,65–67]. The repeats and flanking regions co-operate to give optimal binding, probably by increasing the number of contact points between tau and the microtubule lattice [51,66]; this may be dependent on a flexibility within the binding domain [51,65].

Tau stabilizes microtubules to the depolymerizing effects of low temperatures and anti-microtubule drugs, by modulating their inherent dynamic instability [68,69]. In the presence of tau, the microtubule growth rate increases due to a suppression of the transition frequency from the recovery to the catastrophe phase of growth [68]. Microtubules stabilized by tau aggregate together into bundles of parallel filaments *in vivo* [66,70–73], presumably because the C-terminus of tau containing the microtubule assembly domain is basic [74] and reduces the net anionic charge on microtubules, enabling them to come together in parallel arrays. Full-length tau constructs do not bundle microtubules *in vitro* (unless the tau:tubulin ratio is high [67,75]), whereas fragments comprising the microtubule-binding domain and the flanking regions induce bundling at lower ratios [51,75]. Tau from dogfish erythrocytes only bundles microtubules during their assembly, suggesting it is a co-assembly factor [76]. Addition of the N-terminal domain (which does not bind microtubules) to C-terminal fragments of tau stops bundling [51,67,75], although Kanai *et al.* [77] found that the long N-terminal arm does enhance bundling.

The length of the N-terminal domain determines the microtubule spacing in MAP-induced bundles [45], and tau fragments lacking this domain bundle microtubules more closely than normal [67,75]. Microtubules assembled with native tau display filamentous projections [78], but are not cross-linked *in vitro* [51], suggesting that the N-terminus functions as a projecting arm that keeps microtubules apart [51,75]. The N-terminal inserts that modify the length of the projection arm may be significant in the developmental regulation and neuronal specificity of tau [51]. For instance, tau$_J$ lacks such inserts [59], while HMM tau has a full complement and is restricted to the adult PNS [63,79].

Tau function and regulation by phosphorylation

The function of tau has been investigated by examining the relationship between the time of tau expression and neurite outgrowth and also by modulating its cellular level. The latter approach involves either microinjecting or transfecting non-neuronal cells to overexpress tau or by using tau 'knockout' mice or tau antisense oligodeoxynucleotides to block expression in neuronal cells.

Tau expression is up-regulated in differentiated pheochromocytoma cells (PC12 cells) [80], while the succession of tau_J by the adult isoforms correlates to an increased microtubule stability in these cells [81]. PC12 cells provide a good model system for the biochemical study of neurite outgrowth. They generate neurite-like processes after being induced to differentiate with nerve growth factor (NGF) and the corresponding biochemical changes between proliferating cells and the differentiated, process-bearing cells reveals which factors may be associated with neuritogenesis. A PC12 cell subclone with an enhanced capacity to grow more neurites containing microtubules of increased stability, expresses three times more HMM tau than the parental cell line [82], indicating that different tau isoforms have varying abilities to regulate microtubule stability and neurite growth.

Microinjection of tau [70] or transfection of its cDNA [45,66,77] into non-neuronal cells that do not normally express it, leads to enhanced assembly and bundling of stable microtubules that accumulate in the cells. Transfected insect Sf9 cells typically develop a single neurite-like process containing parallel arrays of microtubules [45,72,73,83]. PC12 cells expressing tau constructs lacking the sequence 154–172 that nucleates microtubules *in vitro* [67], have less stable microtubules compared with cells expressing full-length constructs [84].

Application of tau antisense oligonucleotides to cultured neurons directly after plating, prevents axonal and dendritic differentiation [85,86]. Similarly in PC12 and neuroblastoma cells, the continual presence of antisense from the time of neurite induction allows only the initial growth of neurites, which then stop elongating or retract; these effects are reversible [81,87]. Thus, the elongation of neurites and their attainment of drug-stability in PC12 cells and neuronal differentiation, depends on the prolonged expression of tau; however, one must not forget that tau 'knockout' mice have grossly normal nervous systems [53].

The microtubule assembly properties of tau can be modulated by phosphorylation at different sites (reviewed in [50,88]). Phosphorylation reduces the assembly competence of tau [89], probably by altering the charge distribution on the molecule that is reflected as a transition to a more acidic pI [89,90]. If sites in the microtubule assembly domain are phosphorylated, this could reduce the electrostatic attraction between tau and tubulin and lower the binding affinity [91]. Most adult bovine brain tau is phosphorylated at Ser^{144}, some isoforms are phosphorylated in addition at Ser^{315} and at least two sites, Ser^{144} and Thr^{147}, are developmentally down-regulated [92]. Phosphorylation at a single residue can have distinct consequences on the function of tau. For instance, phosphorylation of Ser^{156} by cyclic AMP-dependent protein kinase reduces the electrophoretic mobility and inhibits microtubule nucleation and growth, interestingly this residue lies in the nucleation domain [67]. Microtubule binding and stability are also suppressed if Ser^{396} [93] and Ser^{262} [94] are phosphorylated, but phosphorylation by Ca^{2+}/calmodulin-dependent kinase does not affect microtubule binding [89].

Tau and microtubule–actin interactions

The first repeat of the microtubule-binding domain of tau binds to actin [95] and recently, neuronal tau and tau-like proteins from non-neuronal cells have been suggested to mediate the interaction between microtubules and F-actin [40,96]. Of great consequence is the possibility that tau-stabilized microtubules in the growth cone interact with filopodial F-actin via an ezrin-like molecule [40,97]. Ezrin is an actin-binding protein associated with the marginal band of erythrocytes [98] and an antibody (13H9) that recognizes it strongly labels actin bundles in the P-domain of growth cones [40,97]. Tau suppression by antisense treatment disrupts F-actin in filopodia and leads to a loss of filopodial 13H9 staining, but does not affect microtubules in the growth cone. Taxol-polymerized microtubules in antisense-treated neurons do not relocate 13H9 staining to the growth cone, suggesting that tau-stabilized microtubules, which under normal conditions lie in close association with the actin bundles, are necessary for the normal localization of the ezrin-like protein [40]. Tau and ezrin thus fulfil some of the requirements to function as a cross-linking mechanism between microtubules and F-actin in growth cones.

MAP 1B

MAP 1B belongs to a two-member group of MAPs, the other member of which is MAP 1A [99,100]. These proteins have molecular masses of 300–350 kDa in SDS/PAGE and distinct immunological and structural properties. MAP1A and MAP 1B are structurally related components of the neuronal cytoskeleton [101]. MAP 1B [102] is also known as MAP 1.2 [103,104], MAP 1X [105,106] and MAP 5 [107].

Developmental expression

Interestingly there is a striking complementary pattern of expression between MAP 1A and MAP 1B during rat brain development [108]. Very early in post-natal development the levels of MAP 1B are at their highest, whereas MAP 1A is hardly detectable. During the second post-natal week, however, the levels of MAP 1B sharply decline [107,109], while the levels of MAP 1A increase to reach a steady basal level which is maintained into adulthood [110]. How does this expression pattern relate to neuronal development? MAP 1B has been suggested to play a crucial role in neurite outgrowth [111], because it is the first MAP to appear in neurites of chick motor neurons and retinal ganglion cells (RGCs) [112,113], and also in the parallel fibres of the rat cerebellum [114]. This view is further substantiated by the fact that MAP 1B is continually expressed in areas of the adult brain which extend axons throughout adult life such as the olfactory epithelium [115], the photosensitive cells of the mammalian retina [116] and also in the mossy fibres of the hippocampus [101]. MAP 1B has been further implicated in neurite outgrowth by experiments using antisense oligodeoxynucleotides to MAP 1B which prevent neuritogenesis in cultured PC12 cells [117].

In contrast to MAP 1B, which appears to be essential for neurite outgrowth, the timing of MAP 1A expression correlates with the establishment of a stable microtubule cytoskeleton more characteristic of the adult nervous system. For instance, when RGCs reach their target neurons in the optic tectum there is an up-regulation of MAP 1A [118,119]. Similarly, during the development of cerebellar Purkinje cells, MAP 1A is initially localized to the cell body and proximal dendrite, but later occurs along the entire dendritic tree [101]. It is also worth noting that in the olfactory bulb, an area which continually expresses MAP 1B, MAP 1A by contrast is absent [101].

This correlative and experimental evidence suggests that the early expression of MAP 1B is essential for the initial events of neuritogenesis, whereas the later expression of MAP 1A relates to a role in neuronal stabilization.

Molecular structure

When MAP 1B was first cloned and sequenced from mouse brain it was reported as 'MAP 1' [120], but the same group have since deduced the complete amino acid sequence of MAP 1B from a series of overlapping genomic and cDNA clones [121]. The encoded protein is 2464 amino acids long, with a molecular mass of 255 534 Da. However, MAP 1B, like tau and intermediate neurofilament proteins (NF-M), is known to run anomalously in SDS/PAGE and has an apparent molecular mass of 325 kDa. MAP 1B has two regions containing repeated motifs. One, located at the C-terminus, is a set of 12 15-amino-acid repeats, separated by two amino acids, while the second is a highly basic N-terminal region with 21 repeats of the motif KKEE or KKEI/V [121]. Subcloned fragments spanning these two distinctive regions were expressed as labelled polypeptides by translation in a cell-free system *in vitro* [121] and then tested for their ability to bind to and co-purify with brain microtubules. Only those peptides with the basic KKEE repeats showed any microtubule-binding ability. To define further the structure of the microtubule-binding domain, full-length and deletion constructs encoding MAP 1B were introduced into cultured HeLa cells by transfection. The expression of transfected polypeptides was monitored by immunofluorescence using antibodies against MAP 1B. These experiments confirmed that MAP 1B binds to microtubules via the basic N-terminal microtubule-binding domain, which is structurally distinct from the microtubule-binding domain of MAP 2 and tau [55,121].

Associated with the microtubule-binding domains of both MAP 1A and MAP 1B are three low-molecular-mass species known as the light chains: LC1 (34 kDa), LC2 (30 kDa) and LC3 (19 kDa) [101,122]. MAP 1A/LC2 and MAP 1B/LC1 are coded by single mRNAs in the same open reading frame and are translated as pre-MAP 1A/LC2 and pre-MAP 1B/LC1 polyprotein precursors that are then proteolytically processed [123,124]. The functional significance of the light chains has not been determined experimentally, but since they bind to microtubules and the heavy chains they may provide structural support by binding to both the heavy chain and the microtubule as a cross-bridge.

The structural results on both MAP 1A and MAP 1B suggests that in developing axons and dendrites the primary function of the molecules is to specifically interact between microtubules. Rotary-shadowed, affinity-purified MAP 1B appears as a long, thin molecule (186 ± 38 nm long) bearing a spherical region at one end which appears to bind to microtubules, suggesting that this structure contains the microtubule-binding domain [125]. The nature of this spherical head domain could be a structural feature of the heavy chain itself or alternatively could represent the heavy chain–light chain complex.

MAP 1B phosphorylation

An intriguing question which is currently being investigated by several laboratories is the functional role of MAP 1B phosphorylation during brain development. MAP 1B migrates as a doublet on SDS/PAGE. The upper band of molecular mass 320–330 kDa appears to be retarded in mobility due to extensive phosphorylation [115,126,127] perhaps specifically on proline-rich sites [128]. There are a number of kinases that are thought to phosphorylate MAP 1B in brain tissue, such as the proline-directed protein kinases (PDPKs) which include cdc2 [129], cdc2-related [130] and mitogen-activated protein kinases (MAP kinases) [131,132]. This hypothesis is based on the fact that monoclonal antibody SMI-31 [133] recognizes phosphorylation-dependent epitopes on MAP 1B, heavy neurofilament proteins (NF-H) and tau [126]. The SMI-31 epitope on both NF-H and tau contains two phosphorylated serines followed by a proline [134]. Furthermore, *in vitro* phosphorylation of recombinant tau with purified PDPKs generates the SMI-31 epitope [134]. Endogenous brain MAP 1B is an *in vitro* substrate for cdc2-like kinases [135,136] which are localized to the growing axons of developing neurons and are strongly down-regulated after neuronal maturation [129].

The functional significance of MAP 1B phosphorylation has been investigated in PC12 cells. Undifferentiated cells normally contain basal levels of largely unphosphorylated MAP 1B which is not associated with microtubules [137]. However, on stimulation with NGF the cells differentiate, sprout neurites and up-regulate MAP 1B [103,104,137,138], in particular the phosphorylated forms [103,104] which bind to the microtubules of growing neurites and their growth cones [137]. Tsao *et al.* [139] have shown that when PC12 cells are stimulated by the addition of NGF or basic fibroblast growth factor, there is a rapid and transient increase in the levels of a serine/threonine kinase designated HMK which phosphorylates high-molecular-mass MAPs and also myelin basic protein *in vitro*. HMK has been shown to be a complex of both extracellular signal-regulated kinases (ERKs) 1 and 2 [140]. For cells to be responsive to NGF they must express the high-affinity NGF receptor, gp140*prototrk* [141,142], which in NGF-stimulated PC12 cells rapidly associates with ERK1 [140]. It is likely that NGF activates an intracellular signal transduction pathway(s) in PC12 cells that is responsible for the phosphorylation of MAP 1B into forms that can bind and stabilize microtubules within the growth cone and growing neurite.

One of the proposed functions for phosphorylated MAP 1B (MAP 1B-P) is the nucleation of microtubules and experiments showing the localization of MAP 1B-P to sites of microtubule nucleation such as the centrosomes of undifferentiated neuroblastoma cells and non-neuronal cells [143,144] are supportive of this idea. It has been shown that at the onset of mitosis in Chinese hamster ovary cells, a class of proteins are expressed that contain a phosphorylated epitope recognized by the monoclonal antibody MPM-2 [145]. Preincubation of isolated centrosomes with MPM-2 or pretreatment with alkaline phosphatase inhibits microtubule nucleation, which suggests that this phosphorylated epitope is necessary for nucleation. Therefore, the quintessential mechanism utilized by these cells to regulate the number of microtubules nucleated by the centrosome on transition from interphase to mitosis may be phosphorylation. The kinase(s) responsible for phosphorylating the MPM-2 site(s) has not been isolated, but a spindle-associated kinase that localizes to the kinetochore fibres, spindle poles and kinetochores has been identified. Interestingly, two of its major endogenous substrates are MAP 4 and MAP 1B [135]. During this G_2/M transition the appearance of the MPM-2 epitope coincides with an increase in maturation-promoting factor [146] which is a complex of p34^{cdc2} kinase and cyclin B (reviewed in [147]). It has also been shown that MAP kinase, a serine/threonine kinase, is also activated downstream of p34^{cdc2} kinase during M phase in *Xenopus* oocytes [148].

By comparing a panel of monoclonal antibodies to phosphorylation-sensitive epitopes on MAP 1B, at least two modes of phosphorylation have been identified [128]. Mode I is classified as the upper band of the doublet which is possibly phosphorylated by PDPKs, whereas mode II exhibits no significant shift in electrophoretic mobility and is proposed to be phosphorylated by casein kinase II (CKII). It has been shown experimentally *in vitro* that MAP 1B phosphorylated by CKII binds microtubules more effectively than unphosphorylated MAP 1B [149,150], and based on the primary sequence of the protein there are 30 potential CKII phosphorylation sites including those flanking the microtubule-binding and projection domains [121]. The *in vitro* dephosphorylation of these two modes is also different and appears to have distinct protein phosphatase (PP) specificities. Mode I sites are effectively dephosphorylated by PP2B, also known as calcineurin, and PP2A; whereas mode II sites are more effectively removed by PP2A and PP1 [151]. It has been proposed that mode I sites are markers for active axonal growth because they are significantly dephosphorylated during development and are barely detectable in adult brain tissue [115,126,128] and also because mode I-phosphorylated MAP 1B is localized to growth cones [20,38,39,152–154]. It is also interesting to note that in the olfactory epithelium, where axonal growth persists, mode I sites remain even in adulthood [152]. In contrast, mode II sites are present in adult brain and also in both axons and dendrites [39,128].

Control of microtubule bundling and actin interactions

The soluble pool of MAP 1B and tau in the growth cone together with their phosphorylation status may play an important role in microtubule bundling and actin interactions. Dephosphorylated tau and mode II-phosphorylated MAP 1B bind more effectively to microtubules [89,150], and tau and mode I-phosphorylated MAP 1B bind to individual microtubules in growth cones [20,40]. Interestingly, inhibition of calcineurin, which dephosphorylates axon-specific isoforms of MAP 1B [151] and tau, blocks axonal growth in cultured cerebellar macroneurons [155]. Significantly, calcineurin is concentrated at the growth cone [155] where growth and steering decisions are made, so it could be here that regulation of MAP phosphorylation is critical.

An extracellular guidance signal transduced through filopodial membrane receptors could involve numerous intracellular cascade events that modify the phosphorylation status of MAPs and actin-binding proteins that regulate the dynamics of the growth cone cytoskeleton. For example, MAP kinases from differentiated PC12 cells phosphorylate MAP 1B *in vitro* [139] and addition of haemolymph reduces the levels of tyrosine phosphorylated proteins at the tips of extending *Aplysia* growth cone filopodia [156]. The combination of MAP 1B and tau phosphorylated to different degrees at various epitopes, may confer microtubules with unique properties required for axonal growth [88]. It may be that the mode I of MAP 1B phosphorylation maintains the microtubules in a dynamic state in the distal motile regions of the growth cone, whereas mode II plays a role in microtubule stabilization, hence its presence in the more stable microtubule bundles in the neurite. Distinct phosphorylated isoforms of tau and MAP 1B become differentially compartmentalized during axonal growth. The PHF-1 phosphorylation epitope on tau is restricted to growing axons of RGCs and axon tracts in the forebrain and is developmentally down-regulated *in vivo* [157]. The tau-1 epitope is restricted to axons, whereas tau phosphorylated at this epitope is localized in the somatodendritic compartment [158]. Mode I- and II-phosphorylated MAP 1B show complementary distributions in growing neurons, with the mode I forms being concentrated towards the growth cone [20,38,39,153,154].

Different developmental expression and phosphorylation patterns of MAP 1A, MAP 1B and tau may determine the plasticity of the microtubule cytoskeleton at different times in neuronal development [108].

The work in our laboratory is funded by the Medical Research Council. M.J. is supported by an Anatomical Society of Great Britain and Northern Ireland Studentship.

References

1. Hynes, R.O. and Lander, A.D. (1992) Cell **68**, 303–322
2. Goodman, C. and Shatz, C. (1993) Cell **72**(Suppl.), 65–75
3. Culotti, J.G. (1994) Curr. Opin. Genet. Dev. **4**, 587–595
4. Tessier-Lavigne, M. (1994) Curr. Opin. Genet. Dev. **4**, 596–601

5.   Kater, S.B. and Rehder, V. (1995) Curr. Opin. Neurobiol. **5**, 68–74
6.   Smith, S.J. (1988) Science **242**, 708–715
7.   Lin, C., Thompson, C.A. and Forscher, P. (1994) Curr. Opin. Neurobiol. **4**, 640–647
8.   Bentley, D. and Toroian-Raymond, A. (1986) Nature (London) **323**, 712–715
9.   Chien, C.-B., Rosenthal, D.E., Harris, W.A. and Holt, C.E. (1993) Neuron **11**, 237–251
10.   Letourneau, P.C. and Shattuck, T.A. (1989) Development **105**, 505–519
11.   Davenport, R.W., Dou, P., Rehder, V. and Kater, S.B. (1993) Nature (London) **361**, 721–723
12.   Sabry, J.H., O'Connor, T.P., Evans, L., Toroian-Raymond, A. and Kirschner, M. (1991) J. Cell Biol. **115**, 381–395
13.   Tanaka, E.M. and Kirschner, M.W. (1991) J. Cell Biol. **115**, 345–363
14.   Tanaka, E. and Kirschner, M.W. (1995) J. Cell Biol. **128**, 127–137
15.   Mitchison, T. and Kirschner, M. (1988) Neuron **1**, 761–772
16.   Gordon-Weeks, P.R. (1993) J. Neurocytol. **22**, 717–725
17.   Tanaka, E., Ho, T. and Kirschner, M.W. (1995) J. Cell Biol. **128**, 139–155
18.   Williamson, T.W., Gordon-Weeks, P.R., Schachner, M. and Taylor, J. (1995) Soc. Neurosci. Abst. **21**, 1775
19.   Gordon-Weeks, P.R. (1991) NeuroReports **2**, 573–576
20.   Bush, M.S., Goold, R., Moya, F. and Gordon-Weeks, P.R. (1996) Eur. J. Neurosci., in the press
21.   Peterson, E.R. and Crain, S.M. (1982) Science **217**, 377–379
22.   Letourneau, P.C. and Ressler, A.H. (1984) J. Cell Biol. **98**, 1355–1362
23.   Letourneau, P.C., Shattuck, T.A. and Ressler, A.H. (1986) J. Neurosci. **6**, 1912–1917
24.   Letourneau, P.C., Shattuck, T.A. and Ressler, A.H. (1987) Cell Motil. Cytoskel. **8**, 193–209
25.   Gordon-Weeks, P.R. (1987) Neuroscience **21**, 977–989
26.   Gordon-Weeks, P.R., Mansfield, S.G. and Curran, I. (1989) Dev. Brain Res. **49**, 305–310
27.   Mansfield, S.G. and Gordon-Weeks, P.R. (1991) J. Neurocytol. **20**, 654–666
28.   Bray, D., Thomas, C. and Shaw, G. (1978) Proc. Natl. Acad. Sci. U.S.A. **75**, 5226–5229
29.   Joshi, H.C., Baas, P., Chu, D.T. and Heidemann, S.R. (1986) Exp. Cell Res. **163**, 233–245
30.   Matus, A., Bernhardt, R., Bodmer, R. and Alaimo, D. (1986) Neuroscience **17**, 371–389
31.   Forscher, P. and Smith, S.J. (1988) J. Cell Biol. **107**, 1505–1516
32.   Lin, C.-H. and Forscher, P. (1993) J. Cell Biol. **121**, 1369–1383
33.   Weisshaar, B., Doll, T. and Matus, A. (1992) Development **116**, 1151–1161
34.   Edson, K., Weisshaar, B. and Matus, A. (1993) Development **117**, 689–700
35.   Heidemann, S.R. and Buxbaum, R.E. (1994) Neurotoxicology **15**, 95–108
36.   Smith, C.L. (1994) J. Neurosci. **14**, 384–398
37.   Bentley, D. and O'Connor, T.P. (1994) Curr. Opin. Neurobiol. **4**, 43–48
38.   Mansfield, S.G., Díaz-Nido, J., Gordon-Weeks, P.R. and Avila, J. (1991) J. Neurocytol. **21**, 1007–1022
39.   Ulloa, L., Diez-Guerra, F.J., Avila, J. and Díaz-Nido, J. (1994) Neuroscience **61**, 211–223
40.   DiTella, M., Fenguin, F., Morfini, G. and Caceres, A. (1994) Cell Motil. Cytoskel. **29**, 117–130
41.   Melki, R., Kerjan, P., Waller, J.P., Carlier, M.F. and Pantaloni, D. (1991) Biochemistry **30**, 11536–11545
42.   Lee, G. and Brandt, R. (1992) Trends Cell Biol. **2**, 286–289
43.   von Massow, A., Mandelkow, E.-M. and Mandelkow, E. (1989) Cell Motil. Cytoskel. **14**, 562–571
44.   Heins, S., Song, Y., Wille, H., Mandelkow, E. and Mandelkow, E.-M. (1991) J. Cell Sci. **14**(Suppl.), 121–124
45.   Chen, J., Kanai, Y., Cowan, N.J. and Hirokawa, N. (1992) Nature (London) **360**, 674–677
46.   Goldberg, D.J., Burmeister, D.W. and Rivas, R.J. (1991) in The Nerve Growth Cone (Letourneau, P.C., Kater, S.B. and Macagno, E.R., eds.), pp. 79–95, Raven Press Ltd, New York
47.   Kuznetsov, S.A., Langford, G.M. and Weiss, D.G. (1992) Nature (London) **356**, 722–725
48.   Taylor, J., Pesheva, P. and Schachner, M. (1993) J. Neurosci. Res. **35**, 347–362
49.   Taylor, J. (1994) in NeuroProtocols (P.R. Gordon-Weeks, ed.), pp. 158–166, Academic Press Inc., San Diego
50.   Lee, G. (1993) Curr. Opin. Cell Biol. **5**, 88–94
51.   Gustke, N., Trinczek, B., Biernat, J., Mandelkow, E.-M. and Mandelkow, E. (1994) Biochemistry **33**, 9511–9522

52. Bush, M.S., Eagles, P.A.M. and Gordon-Weeks, P.R. (1996) in Treatise on the Cytoskeleton, III Cytoskeleton in Specialized Tissues (Hesketh, J.E. and Pryme, I.F., eds.), JAI Press, Connecticut, in the press
53. Harada, A., Oguchi, K., Okabe, S. et al. (1994) Nature (London) **369**, 488–491
54. Takemura, R., Kanai, Y. and Hirokawa, N. (1991) Neuroscience **44**, 393–407
55. Lee, G., Cowan, N. and Kirschner, M. (1988) Science **239**, 285–288
56. Goedert, M., Spillantini, M.G., Jakes, R., Rutherford, D. and Crowther, R.A. (1989) Neuron **3**, 519–526
57. Goedert, M., Spillantini, M.G., Potier, M.C., Ulrich, J. and Crowther, R.A. (1989) EMBO J. **8**, 393–399
58. Goedert, M. and Jakes, R. (1990) EMBO J. **9**, 4225–4230
59. Kosik, K.S., Orecchio, L.D., Bakalis, S. and Neve, R.L. (1989) Neuron **2**, 1389–1397
60. Couchie, D., Mavilia, C., Georgieff, I.S. et al. (1992) Proc. Natl. Acad. Sci. U.S.A. **89**, 4378–4381
61. Goedert, M., Spillantini, M.G. and Crowther, R.A. (1992) Proc. Natl. Acad. Sci. U.S.A. **89**, 1983–1987
62. Georgieff, I.S., Liem, R.K., Mellado, W., Nunez, J. and Shelanski, M.L. (1991) J. Cell Sci. **100**, 55–60
63. Oblinger, M.M., Argasinski, A., Wong, J. and Kosik, K.S. (1991) J. Neurosci. **11**, 2453–2459
64. Lee, G., Neve, R.L. and Kosik, K.S. (1989) Neuron **2**, 1615–1624
65. Butner, K.A. and Kirschner, M.W. (1991) J. Cell Biol. **115**, 717–730
66. Lee, G. and Rook, S.L. (1992) J. Cell Sci. **102**, 227–237
67. Brandt, R. and Lee, G. (1993) J. Biol. Chem. **268**, 3414–3419
68. Drechsel, D.N., Hyman, A.A., Cobb, M.H. and Kirschner, M.W. (1992) Mol. Biol. Cell **3**, 1141–1154
69. Pryer, N.K., Walker, R.A., Skeen, V.P. et al. (1993) J. Cell Sci. **103**, 965–976
70. Drubin, D.G. and Kirschner, M.W. (1986) J. Cell Biol. **103**, 2739–2746
71. Kanai, Y., Takemura, R., Oshima, T. et al. (1989) J. Cell Biol. **109**, 1173–1184
72. Baas, P.W., Pienkowski, T.P. and Kosik, K.S. (1991) J. Cell Biol. **115**, 1333–1344
73. Knops, J., Kosik, K.S., Lee, G. et al. (1991) J. Cell Biol. **114**, 725–733
74. Himmler, A., Drechsel, D., Kirschner, M.W. and Martin, D.W. (1989) Mol. Cell. Biol. **9**, 1381–1388
75. Brandt, R. and Lee, G. (1994) Cell Motil. Cytoskel. **28**, 143–154
76. Sanchez, I. and Cohen, W.D. (1994) Cell Motil. Cytoskel. **29**, 57–71
77. Kanai, Y., Chen, J. and Hirokawa, N. (1992) EMBO J. **11**, 3953–3961
78. Hirokawa, N., Shiomura, Y. and Okabe, S. (1988) J. Cell Biol. **107**, 1449–1459
79. Taleghany, N. and Oblinger, M.M. (1989) J. Neurosci. Res. **33**, 257–265
80. Drubin, D., Kobayashi, S., Kellogg, D. and Kirschner, M. (1988) J. Cell Biol. **106**, 1583–1591
81. Hanemaaijer, R. and Ginzburg, I. (1991) J. Neurosci. Res. **30**, 163–171
82. Teng, K.K., Georgieff, I.S., Aletta, J.M. et al. (1993) J. Cell Sci. **106**, 611–626
83. Knowles, R., LeClerc, N. and Kosik, K.S. (1994) Cell Motil. Cytoskel. **28**, 256–264
84. Léger, J.G., Brandt, R. and Lee, G. (1994) J. Cell Sci. **107**, 3403–3412
85. Caceres, A. and Kosik, K.S. (1990) Nature (London) **343**, 461–463
86. Caceres, A., Potrebic, S. and Kosik, K.S. (1991) J. Neurosci. **11**, 1515–1523
87. Shea, T.B., Beermann, M.L., Nixon, R.A. and Fischer, I. (1992) J. Neurosci. Res. **32**, 363–374
88. Avila, J., Domínguez, J. and Díaz-Nido, J. (1994) Int. J. Dev. Biol. **38**, 13–25
89. Johnson, G.V.W. (1992) J. Neurochem. **59**, 2056–2062
90. Litersky, J.M., Scott, C.W. and Johnson, G.V.W. (1993) Brain Res. **604**, 32–40
91. Correas, I., Diaz-Nido, J. and Avila, J. (1992) J. Biol. Chem. **267**, 15721–15728
92. Arioka, M., Tsukamoto, M., Ishiguro, K. et al. (1993) J. Neurochem. **60**, 461–468
93. Bramblett, G.T., Goedert, M., Jakes, R. et al. (1993) Neuron **10**, 1089–1099
94. Biernat, J., Gustke, N., Drewes, G., Mandelkow, E.-M. and Mandelkow, E. (1993) Neuron **11**, 153–163
95. Moraga, D.M., Nuñez, P., Garrido, J. and Maccioni, R.B. (1993) J. Neurochem. **61**, 979–986
96. Cross, D., Vial, C. and Maccioni, R.B. (1993) J. Cell Sci. **105**, 51–60
97. Goslin, K., Birgbauer, E., Banker, G. and Solomon, F. (1989) J. Cell Biol. **109**, 1621–1631
98. Birgbauer, E. and Solomon, F. (1989) J. Cell Biol. **109**, 1609–1620

99. Bloom, G.S., Schoenfeld, T.A. and Vallee, R.B. (1984) J. Cell Biol. **98**, 320–330
100. Wiche, G., Herrmann, H., Dalton, J.M., Foisner, T., Leichtfried, F.E., Lassmann, H., Koszka, C. and Briones, E. (1986) Ann. N.Y. Acad. Sci. **466**, 180–198
101. Schoenfeld, T.A., McKerracher, L., Obar, R. and Vallee, R.B. (1989) J. Neurosci. **9**, 1712–1730
102. Bloom, G.S., Luca, F.C. and Vallee, R.B. (1985) Proc. Natl. Acad. Sci. U.S.A. **82**, 5404–5408
103. Greene, L.A., Liem, R.K.H. and Shelanski, M.L. (1983) J. Cell Biol. **96**, 76–83
104. Aletta, J.M., Lewis, S.A., Cowan, N.J. and Greene, L.A. (1988) J. Cell Biol. **106**, 1573–1581
105. Binder, L.I., Frankfurter, A., Kim, H., Caceres, A., Payne, M.R. and Rebhun, L.I. (1984) Proc. Natl. Acad. Sci. U.S.A. **81**, 5613–5617
106. Calvert, R. and Anderton, B.H. (1985) EMBO J. **4**, 1171–1176
107. Riederer, B., Cohen, R. and Matus, A. (1986) J. Neurocytol. **15**, 763–775
108. Matus, A. (1988) Annu. Rev. Neurosci. **11**, 29–44
109. Garner, C.C., Garner, A., Huber, G., Kozak, C. and Matus, A. (1990) J. Neurochem. **55**, 146–154
110. Riederer, B. and Matus, A. (1985) Proc. Natl. Acad. Sci. U.S.A. **82**, 6006–6009
111. Tucker, R.P. (1990) Brain Res. Rev. **15**, 101–120
112. Tucker, R.P., Binder, L.I. and Matus, A. (1988) J. Comp. Neurol. **271**, 44–55
113. Tucker, R.P. and Matus, A. (1987) Development **101**, 535–546
114. Calvert, R.A., Woodhams, P.L. and Anderton, B.H. (1987) Neuroscience **23**, 131–141
115. Viereck, C., Tucker, R.P. and Matus, A. (1989) J. Neurosci. **9**, 3547–3557
116. Tucker, R.P. and Matus, A. (1988) Dev. Biol. **130**, 423–434
117. Brugg, B., Reddy, D. and Matus, A. (1993) Neuroscience **52**, 489–496
118. McKerracher, L., Vallee, R.B. and Aguayo, A.J. (1989) Vis. Neurosci. **2**, 349–356
119. Okabe, S., Shiomura, Y. and Hirokawa, N. (1989) Brain Res. **483**, 335–346
120. Lewis, S.A., Sherline, P. and Cowan, N.J. (1986) J. Cell Biol. **102**, 2106–2114
121. Noble, M., Lewis, S.A. and Cowan, N.J. (1989) J. Cell Biol. **109**, 3367–3376
122. Vallee, R.B. and Davis, S.S. (1983) Proc. Natl. Acad. Sci. U.S.A. **80**, 1342–1346
123. Hammarback, J.A., Obar, R.A., Hughes, S.M. and Vallee, R.B. (1991) Neuron **7**, 129–139
124. Langkopf, A., Hammarback, J.A., Muller, R. and Vallee, R.B. (1992) J. Biol. Chem. **267**, 16561–16566
125. Sato-Yoshitake, R., Shiomura, Y., Miyasaka, H. and Hirokawa, N. (1989) Neuron **3**, 229–238
126. Fischer, I. and Romano-Clarke, G. (1990) J. Neurochem. **55**, 328–333
127. Riederer, B., Guadano-Ferraz, A. and Innocenti, G.M. (1990) Dev. Brain Res. **56**, 235–243
128. Ulloa, L., Avila, J. and Diaz-Nido, J. (1993) J. Neurochem. **61**, 961–972
129. Hayes, T.E., Valtz, M.L.M. and McKay, R.D.G. (1991) New Biol. **3**, 259–269
130. Lew, J., Beaudette, K., Litwin, C.M.E. and Wang, J.H. (1992) J. Biol. Chem. **267**, 13383–13390
131. Boulton, T.G., Nye, S.H., Robbins, D.J. et al. (1991) Cell **65**, 663–675
132. Schanen, N.C. and Landreth, G. (1992) Mol. Brain Res. **14**, 43–50
133. Sternberger, L.A. and Sternberger, N.H. (1983) Proc. Natl. Acad. Sci. U.S.A. **80**, 6126–6130
134. Lichtenberg-Kraag, B., Mandelkow, E.-M., Biernat, J. et al. (1992) Proc. Natl. Acad. Sci. U.S.A. **89**, 5384–5388
135. Tombes, R.H., Peloquin, J.G. and Borisy, G.G. (1991) Cell Regul. **2**, 861–874
136. Diaz-Nido, J. and Avila, J. (1992) Second Messenger Phosphoproteins **14**, 39–53
137. Brugg, B. and Matus, A. (1988) J. Cell Biol. **107**, 643–650
138. Drubin, D.G., Feinstein, S.C., Shooter, E.M. and Kirschner, M.W. (1985) J. Cell Biol. **101**, 1799–1807
139. Tsao, H., Aletta, J.M. and Greene, L.A. (1990) J. Biol. Chem. **265**, 15471–15480
140. Loeb, D.M., Tsao, H., Cobb, M.H. and Greene, L.A. (1992) Neuron **9**, 1053–1065
141. Kaplan, D.R., Hempstead, B., Martin-Zanca, D., Chao, M. and Parada, L.F. (1991) Science **252**, 554–557
142. Klein, R., Jing, S., Nanduri, V., O'Rourke, E. and Barbacid, M. (1991) Cell **65**, 189–197
143. Díaz-Nido, J. and Avila, J. (1989) J. Cell Sci. **92**, 607–620
144. Díaz-Nido, J., Armas-Portela, R., Correas, I. et al. (1991) J. Cell Sci. **15**(Suppl.), 51–59
145. Centonze, V.E. and Borisy, G.G. (1990) J. Cell Sci. **95**, 405–411
146. Kishimoto, T. (1988) Dev. Growth Diff. **30**, 105–115
147. Nurse, P. (1990) Nature (London) **344**, 503–508
148. Gotoh, Y., Moriyama, K., Matsuda, S. et al. (1991) EMBO J. **10**, 2661–2668

149. Díaz-Nido, J., Serrano, L., Mendez, E. and Avila, J. (1988) J. Cell Biol. **106**, 2057–2065
150. Ulloa, L., Díaz-Nido, J. and Avila, J. (1993) EMBO J. **12**, 1633–1640
151. Ulloa, L., Dombradi, V., Díaz-Nido, J. et al. (1993) FEBS Lett. **330**, 85–89
152. Gordon-Weeks, P.R., Mansfield, S.G., Alberto, C., Johnstone, M. and Moya, F. (1993) Eur. J. Neurosci. **5**, 1302–1311
153. Black, M.M., Slaughter, T. and Fischer, I. (1994) J. Neurosci. **14**, 857–870
154. Bush, M.S. and Gordon-Weeks, P.R. (1994) J. Neurocytol. **23**, 682–698
155. Ferreira, A., Kincaid, R. and Kosik, K.S. (1993) Mol. Biol. Cell **4**, 1225–1238
156. Wu, D.-Y. and Goldberg, D.J. (1993) J. Cell Biol. **123**, 653–664
157. Pope, W., Enam, S.A., Bawa, N. et al. (1993) Exp. Neurobiol. **120**, 106–113
158. Papasozomenos, S.C. and Binder, J.I. (1987) Cell Motil. Cytoskel. **8**, 210–226

5

Functional domains and intracellular signalling: clues to growth cone dynamics

Roger W. Davenport

Laboratory of Developmental Neurobiology,
National Institute of Child Health and Human Development,
National Institutes of Health, Bethesda, MD 20892, U.S.A.

Introduction

Axons must traverse great distances from their parent cell body and follow tortuous paths to reach their targets. Pathfinding and target recognition during axon extension are critical developmental events involving numerous interactions between separate cells and between cells and the substrate. Such interactions occur at the forefront of neuronal advance, where the growth cone resides and is specialized for such interactions. Identified over 100 years ago, the neuronal growth cone is one of the best-studied organelles in the developing nervous system. Ramon y Cajal [1] was first to observe the ends of developing nerve fibres and this observation truly represented a major breakthrough in understanding neuronal development. Even from his limited observations of fixed preparations, Cajal understood growth cones to be the navigators of neurite extension. Furthermore, he proposed that growth cones may have chemotactic sensitivity that guides them towards targets which serve as sources for diffusible attractive substances. Such chemoattraction of growth cones currently receives considerable attention [2–7]. Clearly Cajal's influence in this field can not be overstated. Together the motor and sensory roles of growth cones that Cajal first imagined are now understood to be fundamental to development of the nervous system.

The dynamic nature of growth cones enables the detection and response to guidance cues, both diffusible and bound, allowing them to fulfill their navigational role. Harrison as the inventor of tissue culture was the first to observe extending fibres and to clearly understand the importance of their activity [8]. Without the ability to observe living growth cones, we would have far less general understanding of growth cone behaviour, of how it is regulated by guidance cues and of what intracellular systems causally underlie its ability to serve as the navigator of neuronal development. Furthermore, the advent of tissue culture allowed a new level of insight: increased resolution and increased accessibility of the growth cone have enlightened us to physical and functional domains within growth cones and how they can be regulated by guidance cues in their environment and intracellular changes in second messenger systems. Though much remains to be unravelled, this review will suggest how three functional domains of the growth cone may be regulated by guidance cues via distinct calcium signalling within the growth cone.

Growth cone structural and functional domains

Neuronal growth cones from numerous species have been observed with a variety of techniques — from preparations that allow one to view growth cones actively growing *in vivo* (e.g. Speidel [9]) and *in vitro* (e.g. Chapters 6, 7 and 10) to fixed preparations that allow one to study their ultrastructure (e.g. Chapters 2 and 4). Growth cones from different cell types have characteristic morphologies at the light and scanning electron microscopy levels [10–12]; however, the anatomy of the growth cone is not a stable fixture. The dynamic morphology of growth cones can most clearly be observed while examining their spontaneous behaviour in culture. Together, such observations allow one to classify three structural domains that comprise growth cones.

Filopodia, like fingers on a hand, represent the most distal domain of the growth cone. Filopodia extend in front and to the side of growth cones from a peripheral domain (P-domain) which is thin and flattened to the substratum. Filopodia are dynamic, thin structures and are perhaps both the best understood and also the most intensively examined structures of growth cones. Their discrete and evanescent nature, their location at the leading edge of the growth cone and their ability to 'reach out' and contact structures within their surroundings have attracted a great deal of experimental attention. Even early investigators noticed filopodia and speculated on their function [8,9,13,14]. Filopodia are filled with a central core of bundles of parallel actin filaments and its regulation remains an area of active investigation (see Chapter 4; [15,16]). Filopodia are very active structures, continuously elongating from and retracting into the growth cone. Photomicrographs definitely do not demonstrate adequately their motility, especially since alongside this filopodial activity, the lamellipodia displays ruffling behaviour and its own (independent?) extension.

Lamellipodia, the thin peripheral margins of the growth cone, comprise the second domain of growth cones. This region consists of a meshwork of actin filaments, and for the most part is void of both microtubules and organelles (see Chapter 4). Recently, however, investigators have demonstrated the dynamic extension of microtubules into this domain (Chapter 4; [17,18]). Endoplasmic reticulum occasionally was observed to be associated with these microtubules, and has been reported to reach the peripheral margins of the growth cone [17,19,20]. Advance of lamellipodia occurs between existing filopodia [21], which can perform a structural role, acting as scaffolding. This form of lamellipodial advance, however, only supports advancement of the P-domain and does not explain overall elongation of the axon. Indeed, ascribing a unique function to this thin lamellae is difficult. Instead, it probably serves in structurally and functionally linking the sensory and motor functions of filopodia to the central regions of the growth cone.

The third domain consists of the thicker, central region of the growth cone. This domain is filled with organelles and is nearly absent of actin filaments [17,22]. The central domain (C-domain) of the growth cone has been suggested to regulate neurite assembly and elongation, because this is the site of microtubule termination (see Chapter 4). Experiments utilizing microtubule-depolymerizing agents [23] as well as experiments demonstrating that microtubules do not move in mass down the neurite [24] have resulted in general support for this proposal. Intuitively, if the C-domain does regulate elongation, then mechanisms located within the lamellipodia must co-ordinate the activity of this domain with the more dynamic, actin-rich domains. This remains an area open for future studies.

Together these three domains help define a contemporary view of a generalized growth cone, based primarily on its form in tissue culture. One sees at the terminal of extending neurites a broad flattened lamellipodium often tipped by numerous filopodia. Both in culture [10,12] and *in situ* [25–27], however, a variety of growth cone shapes are observed. Growth cones vary in their overall size, number and length of filopodia, amount of lamellipodia and overall complexity. Growth cone shape is neither a random event nor is it strictly genetically governed. Studies *in vivo* [11,28–30] and *in vitro* [31–34] convincingly demonstrate that the shape of growth cones and its number of filopodia depend on the environment they traverse, with growth cones exhibiting more complicated morphologies at 'decision points' [35–39]. While the molecular details of processes which precisely regulate growth cone shapes *in vivo* remains uncertain and despite the perplexing variety of growth cone forms, investigations have made progress concerning the effect of a variety of classes of physiological stimuli which affect growth cone morphology.

Stimuli which affect growth cone morphology may be diffusible and dispersed widely or they may be bound to the substratum and encountered only in discrete locations. All primarily affect the growth cone, but may have widespread effects across the entire cell. Growth cone responses can be limited to particular cells types [6,40–42], process types (axons versus dendrites) [43–45], developmental ages (Chapter 8) and can affect subsequent responses of the same growth cone [2,46,47] or even different growth cones entirely across the cell [48] (Fig. 1). Thus, growth cone behaviour may result from both dynamic interactions at their distal extensions and previous manifold interactions within their environment. Clearly each interaction and the resultant behaviour could be under influential regulation by second messenger systems. Indeed, second messengers have been implicated in many of these interactions and their potential regulatory role will be discussed in

Fig. I. **Growth cone behaviour upon encounter with discreet guidance cues is dependent on multiple cellular variables**

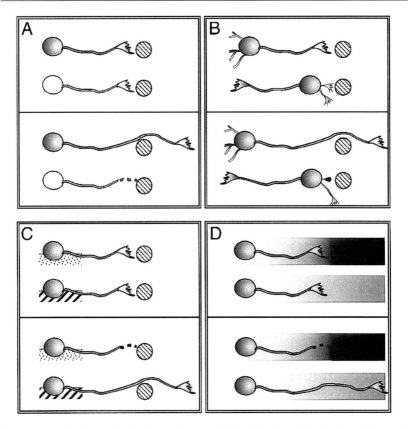

Two cells with their elongating processes and motile growth cones are shown in each upper panel just prior to encounter of guidance cues [indicated by the striped ball in (A)–(C), and as a smooth gradient in (D)].The lower panels indicate resultant growth cone behaviour. Different cell types (A) and even different processes on the same cell type, for example axons versus dendrites (B), respond distinctly to identical guidance cues. Even the same growth cone can respond differently to an encountered stimulus, depending on its developmental age or other environmental cues encountered previously (C) or on the spatial distribution of that guidance cue (D).

several chapters in this volume (Chapters 2, 3, 6, 7 and 10). In the past decade a particular emphasis has been focused on the role of calcium levels within growth cones. Changes in intracellular calcium concentration ($[Ca^{2+}]_i$) upon encounter with guidance cues is discussed in the remainder of this chapter. Throughout this discussion, what must keep be kept in mind is that multiple, interacting molecular cues influence growth cones throughout their lifetime. Thus, although this discussion will focus on the effects of alterations in intracellular calcium levels, it is not presumed this will be to the exclusion of other intracellular changes.

Intracellular calcium and growth cone behaviour

Neuronal growth cones act as navigators during the pathfinding and target recognition critical for normal neuronal development. This role requires growth cones to exhibit a diverse repertoire of behaviours. One fundamental concern of developmental neurobiology is the definition of environmental and intracellular cues that regulate growth cone behaviour. Numerous investigations continue to examine the response of growth cones to a variety of physiological stimuli, applied singly or in combinations (see Chapters 2–4, 6–8 and 10). Intracellular mechanisms responsible for regulating growth cone behaviour have also been addressed. In particular, one second messenger is examined most frequently: intracellular calcium is now understood to serve a key role in regulation of both local and overall growth cone morphology. General evidence from a number of laboratories will be discussed briefly. Later a special emphasis will be placed on the role that changes in intracellular calcium can play in the more subtle changes in growth cone morphology.

Spontaneous fluctuations in calcium and outgrowth

Regulation of neurite elongation by $[Ca^{2+}]_i$ has been actively investigated for nearly 20 years. The number of such studies has multiplied since the development of the calcium indicator fura-2 [49,50]. Calcium indicators allow direct determination of the relationship between $[Ca^{2+}]_i$ and growth cone behaviour. Many studies have been reported in the intervening years and most have demonstrated a general correlation between intracellular calcium levels and growth cone motility. For example, Silver et al. [51] monitored the behaviour of individual growth cones and their spontaneously varying levels of calcium and noted the following correlation: intracellular calcium levels in motile, advancing growth cones were low; higher calcium concentrations were found in motile growth cones that were not advancing, suggesting that a small elevation of calcium inhibits neurite extension. A further rise of intracellular calcium, above the level found in motile non-advancing growth cones, appeared to inhibit motility and cause retraction of growth cones back towards their cell bodies. Some other studies suggested that calcium levels are lower in growth cones that had spontaneously stopped growing [52,53]. These results suggest that in active growth cones changes in calcium levels provide a sufficient condition for regulation of growth cone morphology and behaviour.

Calcium ionophores

The correlation between increases in intracellular calcium and inhibition of growth cone motility can be examined more directly using calcium ionophores. The use of ionophores offers certain experimental advantages; i.e. it produces a uniformly distributed and non-inactivating permeability to calcium. The activity of growth cones of *Helisoma* neurons and hippocampal neurons can be selectively regulated by the addition of the calcium ionophore A23187 [42,54,55]. Exposure to low levels of A23187 can completely inhibit neurite outgrowth, though axonal growth cones of embryonic hippocampal pyramidal neurons were far less sensitive to A23187 than their dendritic counterparts [55]. This indicated that differences in the calcium second messenger system may underlie differences in the ability of growth cones to respond to various guidance cues (Fig. 1). More recently, the growth-inhibitory effects of A23187 were linked to changes in the cytoskeletal machinery underlying neurite elongation [56,57]; filopodia, lamellipodia and their underlying cytoskeleton are sensitive to the effects of A23187 (see Fig. 2 and [54,56,57]). Taken together, these experiments emphasize the regulatory role calcium can serve in controlling overall neuronal growth cone shape and function.

Neurotransmitters and electrical activity

Results from experiments utilizing physiological stimuli such as neurotransmitters and electrical activity both support the idea that rises in intracellular calcium levels are inhibitory to growth cone motility. Work with neurotransmitters provided the first direct evidence that rises in $[Ca^{2+}]_i$ inhibit growth cone motility. Direct measurements of $[Ca^{2+}]_i$ within growth cones of particular, identified *Helisoma* neurons revealed that addition of excitatory neurotransmitters produced a neuron-specific large rise in calcium levels and inhibition of outgrowth [41,58,59]. The selective effect of neurotransmitters on $[Ca^{2+}]_i$ and growth cone motility provide just one piece of evidence that intracellular calcium levels could underlie some cell-specific responses to guidance cues (Fig. 1). Similarly, the excitatory neurotransmitter glutamate raised $[Ca^{2+}]_i$ and immobilized growth cones on dendrites of cultured hippocampal pyramidal neurons [43,44,60]. The presence of the inhibitory neurotransmitter γ-aminobutyric acid, along with its potentiator diazepam, blocks the calcium rise associated with glutamate application as well as the dendritic regression. The effects of neurotransmitters appear to be mediated through changes in membrane potential [61] and thus perhaps it is not surprising that action potentials themselves can produce a rise in intracellular calcium levels and inhibition of outgrowth [58,62–64]. In *Helisoma* neurons, controlling membrane potential while simultaneously applying excitatory neurotransmitters can override the potential growth inhibition of the neurotransmitter stimuli [61]. Together, these findings suggest that electrical activity, whether mediated directly or through neurotransmitters, influences neuronal outgrowth via changes in $[Ca^{2+}]_i$. This is especially relevant since activity-dependent processes play a prominent role during development and regeneration of the nervous system [65,66].

Fig. 2.   **Increases in calcium influx dramatically alter filopodial disposition and lamellipodial extent**

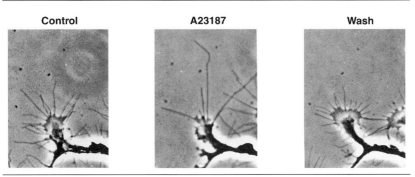

| Control | A23187 | Wash |

A *Helisoma* growth cone is shown before, during and after application of the calcium ionophore A23187. Filopodia elongate, but many are simultaneously lost (i.e. several filopodia get much longer while the rest retract); lamellipodia also withdraws in the presence of A23187, an agent which directly increases calcium influx. The effects on filopodia and [Ca²⁺]ᵢ were quantified and highly correlated with the magnitude of the rise in [Ca²⁺]ᵢ [54] and are reversible ('wash').

Contact-dependent guidance cues

Contact-dependent guidance may underlie the majority of neuronal development, but far less is known about the effect of bound molecules on intracellular calcium levels within growth cones. In some systems, calcium measurements during growth cone collapse mediated by cell–cell contact show no change in $[Ca^{2+}]_i$ [67], indicating that alternative, calcium-independent pathways for evoking growth cone collapse probably do exist. However, a number of independent and very interesting results concerning the mediating effects of intracellular calcium levels in response to growth-promoting substrates and growth-inhibiting guidance molecules have been reported recently. These are discussed more completely in separate chapters of this volume (see Chapters 3, 6 and 7).

Growth cone calcium levels: a continuum and an optimum

Kater and colleagues [68,69] suggested early on that control of neuronal outgrowth directly involved intracellular calcium. Their model proposed that growth cone activity and outgrowth occur over a range of calcium concentrations, with maximal activity occurring at an optimal level within this generally permissive range (Fig. 3). If calcium levels were to change too far from the optimum, moving outside the permissive range, then growth cones would collapse and neurite outgrowth would cease. Thus, depending on calcium levels at rest, a given stimulus could have opposite effects. For example, with a neuron that is growing slowly as a result of permissive but suboptimal levels of intracellular calcium, a given stimulus could raise intracellular calcium closer to the optimal level and, accordingly, stimulate neurite outgrowth. On the other hand, with a neuron that is already growing at a maximal rate, the same stimulus could raise $[Ca^{2+}]_i$ above the permissive range and inhibit motility. This model may explain why both ionophore and electrical depolarization [64,70–76] can both promote and inhibit neurite outgrowth.

Fig. 3.   **Neuronal outgrowth is affected by changes in $[Ca^{2+}]_i$, in large part dependent upon the direction and magnitude of the change**

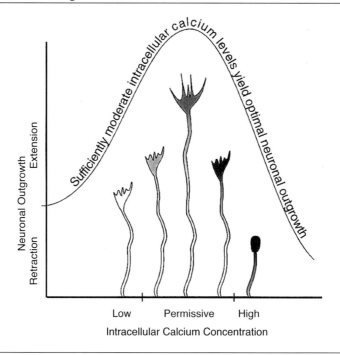

A model for calcium regulation of neuronal outgrowth suggests that neurite outgrowth and growth cone motility are optimal only at moderate $[Ca^{2+}]_i$. Increasing or decreasing intracellular calcium from the optimum results in decreased outgrowth and motility. Low $[Ca^{2+}]_i$ are associated with immobilization, whereas increasing intracellular calcium beyond the permissive range can lead to complete growth cone collapse and neurite retraction.

At least two investigations on two different cell types have now reported direct demonstrations of a dual sensitivity of growth cones to changes in intracellular calcium levels. In one study [77], neurite outgrowth in rat dorsal root ganglion (DRG) cells was examined over a range of $[Ca^{2+}]_i$. Calcium levels were manipulated by addition of a calcium chelator, 1,2-bis(o-aminophenoxy)ethane-$NNN'N'$-tetraacetic acid (BAPTA; [78]), to the culture medium at the time of plating; $[Ca^{2+}]_i$ was assessed after 24 h in culture by loading the cells with fura-2. The fraction of cells with regrowing neurites, as well as the length of the longest neurite on each neurofilament-positive cell, were measured. Neurite outgrowth, by both criteria, was maximal at intermediate calcium levels and was reduced at higher and lower values of $[Ca^{2+}]_i$. These results provide direct evidence for the suggestion that an optimal calcium range exists for neurite outgrowth.

A second investigation on the dual effect which calcium levels exert on neuronal growth cones examined the quantitative changes in filopodia on *Helisoma* growth cones [54]. Experimentally evoked changes in intracellular calcium levels were measured and directly correlated with growth cone filopodial morphology. A rise in $[Ca^{2+}]_i$ caused filopodia first to elongate, then later to retract (see Fig. 2). The magnitude of both filopodial elongation and filopodial loss correlated well with the transient peak values of $[Ca^{2+}]_i$ reached during a given experimental treatment [54]. Thus, it was concluded that filopodial numbers are not determined by the absolute value of intracellular calcium present at any given time, but that the magnitude of filopodial loss is determined by the magnitude of the initial calcium rise. Furthermore, the precise effect of a stimulus, whether especially increasing filopodial length or decreasing filopodial numbers, also appears dependent upon the magnitude of the initial calcium rise.

Both studies support the working model that intracellular calcium levels can exert a stimulatory or an inhibitory response for neuronal outgrowth and the exploratory behaviour of growth cones, depending on both the magnitude of the change in calcium levels and the existing basal calcium levels. Clearly this model excludes complexities of the intracellular environment which also may affect growth cone behaviour. For example, Fields and co-workers [63,64] demonstrated that the rate of change, rather than the maximal calcium level, is a better predictor for the response of DRG growth cones to electrical stimulation. Indeed, very large but relatively rapid transients in $[Ca^{2+}]_i$ have been used to elicit filopodial production and elongation, rather than inhibition (for example, see Figs. 4 and 5). Moreover, as discussed in more detail later, a rise in $[Ca^{2+}]_i$ can be spatially hetero-geneous across the growth cone, as a result of the clustering of calcium channels [79] or non-uniform stimulation [80,81]. Such local rises in $[Ca^{2+}]_i$ may not be reflected nor perhaps related in a simple manner to overall $[Ca^{2+}]_i$ or motility across the growth cone. The working model has provided a useful platform over the past 7 years from which to assess a variety of phenomena relating to growth cone behaviour. Today, it can serve as a foundation upon which we can examine results which probe regulation of growth cone behaviour with even higher resolution.

Control of growth cone turning and more regional and even quite local growth cone behaviours represent one of the most exciting topics related to regulation of growth cone behaviour. Directional neuronal guidance lies at the heart of our interest in growth cones and indeed nearly every chapter in this volume addresses this topic. Local changes in $[Ca^{2+}]_i$ have been associated with local changes in growth cone morphology and these results will be discussed next. The most exciting interpretation is that the extent of intracellular signalling could account for the diverse and navigational behaviour of growth cones upon encounter with developmental guidance cues. Thus, the rest of this chapter will elaborate more extensively on recent results which provide a novel dissection of the underlying calcium second messenger architecture which subserves growth cone motility.

Dissection of growth cone signalling

A number of studies have implicated calcium as a regulator of discreet changes in growth cone morphology. Goldberg [82] evoked local lamellipodial extension from *Aplysia* growth cones by focally applying an extracellular calcium source near a growth cone bathed in calcium-free medium and thus suspected changes in $[Ca^{2+}]_i$ to be the causal second messenger. Several investigators have applied focal electric fields to orient neurite outgrowth of embryonic neurons from the dissociated neural tube of *Xenopus* (see Chapter 10). McCaig [83] demonstrated an increase in filopodial number only on the cathode side of growth cones, concomitant with orientation changes. With inorganic calcium-channel blockers, the change in filopodial distribution on *Xenopus* growth cones was prevented [84]. McCaig and others have suggested that electric fields may mediate their guidance effects via changes in $[Ca^{2+}]_i$ [80,81,85,86]. The effect which local changes in $[Ca^{2+}]_i$ have on local growth cone morphology has now been addressed more directly. Recent experiments, some previously published and some not, which directly examine calcium signals within growth cones and their determination of regional and even very localized changes in growth cone morphology are discussed.

Hotspots
One of the most simplistic methods to produce regional changes in $[Ca^{2+}]_i$ and growth cone morphology is to directly, physically stimulate the growth cone. Wessells and Nuttall [87] directly lifted edges of motile growth cones and observed that subsequent outgrowth was directed away, as if adhesion to the surface was providing a directional influence. Mills *et al.* initially combined mechanical perturbation of growth cones with calcium imaging [88]. Both $[Ca^{2+}]_i$ and filopodial distribution could be regionally affected by mechanical aberration (R.W. Davenport, L.R. Mills and S.B. Kater, unpublished work), as indicated in Fig. 4. Whereas the causal link to either the change in morphology or $[Ca^{2+}]_i$ remains difficult to discriminate, these studies indicate a strong and dramatic correlation between regional changes in calcium levels and growth cone behaviour. Thereby, such seemingly simplistic correlations provide an exciting link to the potentially important developmental role of local changes in the calcium second messenger system.

Fig. 4.   **Mechanical disturbance of motile growth cones can lead to dramatic changes in both local [Ca²⁺]ᵢ and filopodial disposition.**

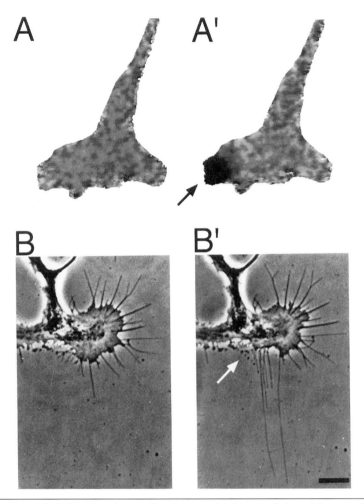

*Simple contact of a Helisoma growth cone with a fine glass micropipette results in spatially restricted changes, i.e. fura-2-loaded growth cones show a local change in [Ca²⁺]ᵢ (**A**). Calcium levels are indicated by shades of grey; high [Ca²⁺]ᵢ are dark. Note the local, high [Ca²⁺]ᵢ (≈μM) region near the site of mechanical disturbance (arrow in **A'**). Motile growth cones display exaggerated filopodial changes that are also spatially restricted to the same degree (**B'**). The ability of rises in intracellular calcium to be localized to subdomains within the growth cone affords the possibility of local control of the motile apparatus. Scale bar = 1 0 μm (R.W. Davenport, L.R. Mills and S.B. Kater, unpublished work).*

Localization of two types of calcium channels were investigated on N1E-115 neuroblastoma cells and indicate that clusters of L-type calcium channels caused $[Ca^{2+}]_i$ hotspots in neuronal growth cones [79]. During 1 s voltage-clamp depolarizations (only growth cones close to the cell body were examined to ensure voltage-clamped conditions), intracellular calcium changes on the order of 1 μM were recorded at the hotspot. Even a single action potential could raise intracellular calcium by nearly 100 nM. In five out of seven growth cones that grew in the 4 min following stimulation, the site at which the greatest outgrowth occurred was close to a hotspot, suggesting a spatial correlation between hotspots and the site of outgrowth. Calcium hotspots produced by clustering of L-type calcium channels therefore may provide a mechanism for triggering local morphological changes and could serve a crucial role in development.

Gradients

Two investigations simultaneously reported [80,81] local changes induced experimentally in growth cone calcium currents. Both investigations took advantage of the high degree of experimental tractability afforded by electric field stimulation paradigms (see Chapter 10). The simplicity of the experimental set up, the ability to quantify stimulus strength and to spatially and temporally localize a stimulus to discrete regions of the growth cone make this paradigm unique. In both studies, electric fields produced graded effects across growth cones, both in terms of outgrowth and calcium concentration.

The study by Bedlack *et al.* [80] demonstrated that N1E-115 neuroblastoma cells turned toward a cathode stimulus in a broad electric field produced across the entire culture dish. Simultaneously, a calcium gradient was produced across these cells. Addition of various calcium-channel blockers, including dihydropyridines which block neuronal L-type calcium channels, prevented both the rise in calcium and the redirection of outgrowth. This suggested a causal relationship between changes in calcium and growth cone navigation and extended the findings of Silver *et al.* [79], which demonstrated a correlation between local 'hotspots' of calcium influx through L-type calcium channels and rapid local morphological changes on these same growth cones.

A second investigation asked whether neuronal growth cones from *Helisoma*, previously studied in the context of global changes in $[Ca^{2+}]_i$ and growth cone behaviour, could display local changes in $[Ca^{2+}]_i$ and whether such local changes also result in local changes in filopodial disposition [81]. In this study, electric fields were applied focally to present a precisely localized and quantified stimulus to active growth cones. Two regional responses of these growth cones were observed: (1) a rise in $[Ca^{2+}]_i$, which begins in a limited region of the growth cone nearest the cathode and then spreads across the growth cone (Fig. 5); and (2) the generation and elongation of filopodia on the side of the growth cone nearest the cathode which coincides with the area of the initial increase in $[Ca^{2+}]_i$. Multiple lines of evidence suggest that changes in $[Ca^{2+}]_i$ indeed were causal to the changes in growth cone morphology.

Fig. 5.	**Focal electric field stimulation can elicit changes in growth cone calcium levels that spread as a gradient across the growth cone**

(a)

(b)

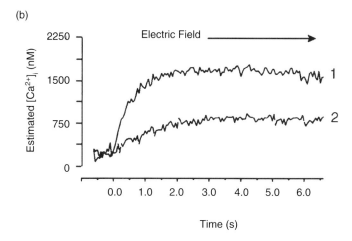

Application of a highly localized electric field to a portion of a Helisoma neuronal growth cone results in a localized rise in $[Ca^{2+}]_i$, as indicated by intracellular fura-2 (a). The rapid increase in $[Ca^{2+}]_i$ and subsequent development of a calcium gradient is shown in (b). Plotted is the estimated $[Ca^{2+}]_i$ from the ratio of the fluorescent emission (F_{pre}/F, 380 nm absorbance) at two separable locations, $= 10 \mu m$ apart for the growth cone shown in (a). The rate of rise was greatest at the site closest to the cathode. Scale bar $= 10 \mu m$.

(1) Induced changes in $[Ca^{2+}]_i$ occurred prior to and with a lower threshold than the induced changes in filopodia. The rise in $[Ca^{2+}]_i$ occurred rapidly (within

milliseconds), but remained highly elevated (greater than twice rest levels) for durations longer than the initial filopodial changes.

(2) Both the initial change in $[Ca^{2+}]_i$ and in filopodia were localized consistently to the cathode side of the growth cone and could be spatially mapped. Each demonstrated a graded response across the growth cone that could be qualitatively and quantitatively recorded.

(3) Induced changes in $[Ca^{2+}]_i$ and in filopodia could be spatially restricted to a similar degree ($\leq 20\%$ of the growth cone) and to a similar location (the site closest to the cathodal source) even when repeating stimulation at different sites on the same growth cone.

(4) Preventing changes in $[Ca^{2+}]_i$ by removing extracellular calcium from the medium, blocked the change in filopodia.

(5) Using a stimulus duration sufficient to cause the greatest change in $[Ca^{2+}]_i$ and the greatest gradient of $[Ca^{2+}]_i$ caused a significant and long-lasting change in filopodia.

(6) For sustained stimulation, the peak $[Ca^{2+}]_i$ reached during the rapid, large and spatially localized rise in $[Ca^{2+}]_i$ was well correlated with local changes in both filopodial length and number.

Taken together, the results strongly implicate a causal relationship between local $[Ca^{2+}]_i$ and local morphological responses of growth cones. Local changes in the morphology of individual growth cones could affect dramatically the final morphology of parent neurons and thereby resultant circuitry. Thus, the demonstration that local changes can occur in the growth cone intracellular calcium second messenger system and that these local changes elicit local changes in the growth cone morphology suggests a role for intracellular calcium signals in 'subtle' aspects of the development of the nervous system.

Filopodia interact with their environment via calcium

Filopodia represent the 'subtle' aspect of growth cone morphology most relevant to neuronal guidance. These dynamic extensions continuously elongate and retract, pulling the growth cone along. The importance of single filopodia to overall neuronal development of the grasshopper limb bud serves as one example to demonstrate a general phenomenon of neuronal development and regeneration: local changes in growth cone behaviour play a crucial role in neuronal guidance. Within the grasshopper limb bud, individual filopodia can direct outgrowth of pioneer neurons by singularly contacting guidance cues and initiating changes that completely alter the path of outgrowth [89]. Several investigators have described different situations in which individual filopodia strongly affect the status of entire growth cones; for example, filopodial contact with a change of substrate is sufficient to redirect or even collapse a growth cone (see Chapters 6 and 7). Numerous such examples demonstrate that local growth cone behaviour, such as selective filopodial extension, alter growth cone navigation and thus can directly affect the final morphology of individual neurons and ultimately the entire neural circuit.

Filopodia appear to be well suited to serve as both antennae-like sensors and mechanical supports during neurite extension. The broad span of filopodia allows sampling of information over a greatly enhanced radius and furthermore, forward projecting filopodia encounter potential cues in their environment long before the advancing growth cone proper. By making use of cell culture, previous studies have demonstrated the ability of filopodia to serve structural roles [21], exert mechanical tension [90–92] and selectively adhere to their surroundings [93–96]. Furthermore, as described in Chapter 2 phosphotyrosine and other developmentally relevant molecules are concentrated at the tips of filopodia. Direct tests of a general and independent sensory and motor role for filopodia were made recently by surgically isolating individual filopodia from their parent growth cone and optically monitoring their morphology and calcium second messenger systems [97].

Isolated filopodia provided the possibility for a novel examination of the independent capabilities of filopodia and to resolve definitively the responsiveness of the filopodia versus the growth cone itself. In that study, neurons were injected with the calcium indicator fura-2 and filopodia subsequently isolated from their parent growth cones by careful transection with a glass micropipette (Fig. 6). Isolated filopodial membranes re-sealed, as indicated by their continued fluorescence, and maintained their elongated form. When stained with rhodamine-phalloidin, which brightly labels filamentous actin (F-actin), no differences in actin distribution were observed when compared with attached filopodia. Furthermore, $[Ca^{2+}]_i$ in isolated filopodia were indistinguishable from those within the growth cone itself. The responsiveness of such isolated filopodia to stimuli known to alter growth cone $[Ca^{2+}]_i$ and behaviour were then examined.

Investigations on several types of neuronal growth cones from different species have shown that agents causing depolarization can increase $[Ca^{2+}]_i$ within the body of growth cones by increasing calcium influx [58,79,80,98]. For example, direct depolarization via an electric field can cause growth cones to completely change behaviour (see Chapter 10), and in some cases this results from an increased $[Ca^{2+}]_i$ [80,81]. Electric fields caused isolated filopodia to raise $[Ca^{2+}]_i$ significantly (Fig. 6); when the stimulus terminated, $[Ca^{2+}]_i$ in filopodia returned to rest values. Given previous demonstrations of voltage-dependent channels on both the somata and growth cones of these neurons [99] and that responses to electric fields in other systems were previously accounted for by the presence of voltage-dependent channels on growth cones [80], such experiments on isolated filopodia indicated the likely presence of these channels on filopodia as well. Moreover, the fact that $[Ca^{2+}]_i$ was restored after termination of the stimulus indicates additionally the presence of clearance and/or buffering mechanisms in filopodia as part of a homoeostatic machinery to maintain filopodia at basal rest levels. Thus, filopodia possess the capability of independently responding to changes in membrane polarization, enabling a coupling of external stimuli to changes in internal messengers.

Fig. 6.   **Surgical isolation of filopodia from motile growth cones**
                reveals their ability to independently respond to stimuli

(a) Load Fura-2 and Isolate Filopodia

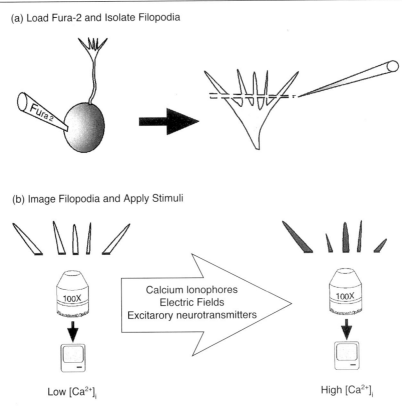

(b) Image Filopodia and Apply Stimuli

Calcium Ionophores
Electric Fields
Excitatory neurotransmitters

100X

Low [Ca²⁺]ᵢ

100X

High [Ca²⁺]ᵢ

(a) Neuronal somata were first filled via pressure injection with fura-2. Subsequently, filopodia were carefully transected from their parent growth cone using a fine glass micropipette, drawn across the filopodia near their base. *(b)* Using high-resolution optics and a computer-based imaging system, intracellular calcium levels were recorded at rest and following application of stimuli known to alter growth cone [Ca²⁺]ᵢ and behaviour. Resting calcium levels were indistinguishable between isolated filopodia and their parent growth cones. Upon stimulation large changes in [Ca²⁺]ᵢ were recorded from isolated filopodia, indicating their independent ability to detect and respond to environmental cues.

Excitatory neurotransmitters are another class of stimuli known to depolarize, raise [Ca²⁺]ᵢ and inhibit motility of growth cones (see previous section, Neuro-transmitters and electrical activity). Addition of dopamine or 5-hydroxytryptamine (serotonin), for example, results in a significant rise in *Helisoma* growth cone [Ca²⁺]ᵢ and an accompanying inhibition of outgrowth [41,58]. Exposure of isolated filopodia to either dopamine or 5-hydroxytryptamine also resulted in a large rise in $[Ca^{2+}]_i$ (Fig. 6). This effect was entirely dependent upon the presence of extracellular calcium and could be negated when acetylcholine was added either before or after. Thus, filopodia can respond independently with an influx of calcium to physiological agents such as neurotransmitters, and furthermore can respond to multiple stimuli in a manner which allows direct signalling to the rest of the growth cone.

The high surface area to volume ratio of filopodia seems to confer an extremely high sensitivity to environmental influences. For example, when the calcium ionophore 4-bromo-A23187 was applied to uniformly increase calcium influx [100], the resultant rise in $[Ca^{2+}]_i$ was significantly greater in isolated filopodia than in their parent growth cones. In fact, assuming a filopodial volume of approx. 1 fl, on the order of only 100 calcium ions could raise $[Ca^{2+}]_i$ by 100 nM. Calcium indicators may not even be sensitive enough to detect potentially relevant $[Ca^{2+}]_i$ changes in structures possessing as little as one-quadrillionth part of a litre of cytoplasm. These results demonstrate what is potentially the most important advantage of relegating a sensory function to the filopodia, namely, the increased sensitivity afforded by the purely physical dimensions of these structures.

To determine whether filopodia also possess independent contractile capabilities the ability of isolated filopodia to change length in response to neurotransmitters was tested. Application of excitatory neurotransmitters significantly decreased the length of isolated filopodia in a dose-dependent manner (Fig. 6). These results demonstrate that filopodia can act as autonomous mechanical units as well as convey information to the growth cone itself.

In summary, filopodia can be considered both as distant extensions of the receptive capacities of growth cones and as autonomous responsive units. Filopodia possess several unique properties that allow them to act as efficient sensory and mechanical probes: filopodia are small, motile and transient structures which probably require much less energy to extend than the larger mass of the growth cone; filopodia have a large surface area to volume ratio compared with growth cones, which permits a high sensitivity to environmental signals; and filopodia are discrete entities extending in a fan-like fashion across the front of the growth cone, which allows growth cones to sample information from many different areas simultaneously and discretely, thus offering the possibility of comparing inputs from across a broad radius. Clearly, growth cone filopodia contain the necessary machinery with which to both sense and respond to environmental cues as well as convey information to the growth cone itself; thus filopodia should be considered as fundamental units of organization for neuronal pathfinding.

Fig. 7.   **Encounters with guidance cues may elicit intracellular signals**
that differentially affect the functional domains at the growth
cone

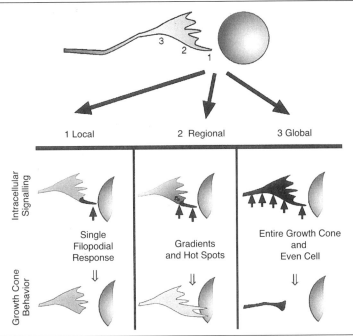

This Figure represents a model which suggests that the spatial extent of second messenger signalling will determine the spatial extent of resultant growth cone behaviour. A number of studies indicate that a primary response of growth cones to particular guidance cues is a rapid change in intracellular second messenger levels, followed by morphological changes. A growth cone's response is not spread uniformly across the growth cone, however. Growth cones respond to guidance cues by very local changes in morphology, such as local filopodial withdrawal; by more regional events, such as turning away or towards an encountered stimulus; and certain stimuli seem to have a powerful effect resulting in extensive changes in growth cone morphology (such as complete collapse) that can affect other growth cones on the same cell. Spatially restricted second messenger changes may underlie the different growth cone behaviours: isolated filopodia can independently regulate their $[Ca^{2+}]_i$ and morphology, suggesting that individual filopodia on a motile growth cone might be able to respond to encountered guidance cues independently; gradients and hotspots of calcium can be induced and could represent the initial spread of intracellular signalling which would then lead to moderate changes in growth cone morphology; extensive or even very global changes in second messengers have been causally linked with growth cone collapse, neurite retraction and changes in outgrowth of even the most distal extensions on the neuron. Future investigation necessarily will be directed at confirming the link between intracellular signalling and growth cone behaviour and revealing determinants of their respective spatial extent.

Co-ordination of filopodia and growth cone behaviour: future directions

Filopodia play a critical role during development of the nervous system, enabling much of what we associate generally with growth cone function to occur. The rest of the growth cone must then co-ordinate the multifarious signals impinging from its complement of filopodia. The demonstration of individual filopodial responses to changes in their immediate environment suggests a very local system of spatial discrimination and responsiveness exists within neuronal growth cones (Fig. 7). Indeed, the very unique abilities of filopodia and their wide expanse in front of the growth cone itself illustrates the requirement for an intracellular system to co-ordinate growth cone behaviour. While individual filopodia situated at the leading edge of growth cones are the first structures to interact with environmental stimuli and serve well in the primary detection and response to external guidance cues, signals from individual filopodia must be co-ordinated to enable directed guidance of the growth cone. This fact is confirmed by results demonstrating that in some cases other factors can override the normally directive role of filopodia, allowing growth cones to turn in a direction away from increased filopodial activity (see Chapter 10). Thus, growth cones must serve both as an integrator of cues from the full span of filopodia and as the site of organization of motility for the elongating neurite. Only by co-ordination of these different components can such processes culminate in successful navigation.

As these results illustrate, intracellular calcium can serve as a regulator of growth cone morphology at even the most refined level. Calcium levels have been directly or indirectly related to overall phenomena such as neurite elongation rate, growth cone turning, collapse and to regional and even quite local changes in morphology such as filopodial and lamellipodial extension. Only a portion of these effects are summarized in Fig. 7. How the calcium second messenger system regulates these effects and how it integrates with other second messenger systems to enable successful outgrowth remains an exciting area for future study. Through this chapter I have attempted to elucidate that a major direction for developmental neurobiology today is towards a better understanding of aspects of intracellular communication which fully co-ordinate growth cone behaviour.

I am thankful for the guidance and direction given to me by my mentors S.B. Kater, F. Bonhoeffer and P.G. Nelson and indebted to the assistance and support of my friend and colleague Edda Thies.

References

1.   Ramon y Cajal, S. (1890) Anat. Anz. **5**, 609–613
2.   Hankin, M. and Lund, R. (1991) Trends Neurosci. **14**, 224–228
3.   Lund, R. and Hankin, M. (1995) J. Comp. Neurol. **356**, 481–489
4.   Dodd, J. and Schuchardt, A. (1995) Cell **81**, 471–474

5. Tessier-Lavigne, M., Placzek, M., Lumsden, A.G.S., Dodd, J. and Jessell, T.M. (1988) Nature (London) **336**, 775–778
6. Colamarino, S. and Tessier-Lavigne, M. (1995) Cell **81**, 621–629
7. Tessier-Lavigne, M. (1994) Curr. Opin. Genet. Dev. **4**, 596–601
8. Harrison, R.G. (1910) J. Exp. Zool. **9**, 787–848
9. Speidel, C.C. (1933) Am. J. Anat. **52**, 1–79
10. Haydon, P.G., Cohan, C.S., McCobb, D.P., Miller, H.R. and Kater, S.B. (1985) Neuron **1**, 919–927
11. Roberts, A. and Taylor, J.S.H. (1983) J. Embryol. Exp. Morphol. **75**, 31–47
12. Roberts, A. and Patton, D.T. (1985) J. Neurosci. Res. **13**, 23–38
13. Weiss, P. (1934) J. Exp. Zool. **68**, 393–448
14. Hughes, A. (1953) J. Anat. **87**, 150–162
15. Bentley, D. and O'Connor, T.P. (1994) Curr. Opin. Neurobiol. **4**, 43–48
16. Lin, C. and Forscher, P. (1995) Neuron **14**, 763–771
17. Dailey, M.E. and Bridgman, P.C. (1989) J. Neurosci. **9**, 1897–1909
18. Lin, C.H. and Forscher, P. (1993) J. Cell Biol. **121**, 1369–1383
19. Kater, S., Dou, P., Mills, L. and Davenport, R. (1994) Soc. Neurosci. Abstr. **20**, 1476
20. Ouyang, Y., Airey, J., Sutko, J. and Ellisman, M. (1994) Soc. Neurosci. Abstr. **20**, 1082
21. Goldberg, D.J. and Burmeister, D.W. (1986) J. Cell Biol. **103**, 1921–1931
22. Forscher, P. and Smith, S.J. (1988) J. Cell Biol. **107**, 1505–1516
23. Bamburg, J.R., Bray, D. and Chapman, K. (1986) Nature (London) **321**, 788–790
24. Lim, S.S., Edson, K.J., Letourneau, P.C. and Borisy, G.G. (1990) J. Cell Biol. **111**, 123–130
25. Bentley, D. and Toroian, R.A. (1986) Nature (London) **323**, 712–715
26. Harris, W., Holt, C. and Bonhoeffer, F. (1987) Development **101**, 123–133
27. Godement, P., Salaun, J. and Mason, C.A. (1990) Neuron **5**, 173–186
28. Taylor, J.S.H. and Roberts, A. (1983) J. Embryol. Exp. Morphol. **75**, 49–66
29. Tosney, K.W. and Landmesser, L.T. (1985) J. Neurosci. **5**, 2345–2358
30. Caudy, M. and Bentley, D. (1986) J. Neurosci. **6**, 364–379
31. Luduena, M.A. (1973) Dev. Biol. **33**, 268–284
32. Letourneau, P.C. (1979) Exp. Cell Res. **124**, 127–138
33. Bray, D. and Chapman, K. (1985) J. Neurosci. **5**, 3204–3213
34. Gomez, T. and Letourneau, P. (1994) J. Neurosci. **14**, 5959–5972
35. Bovolenta, P. and Mason, C. (1987) J. Neurosci. **7**, 1447–1460
36. Taghert, P.H., Bastiani, M.J., Ho, R.K. and Goodman, C.S. (1982) Dev. Biol. **94**, 391–399
37. Raper, J.A., Bastiani, M. and Goodman, C.S. (1983) J. Neurosci. **3**, 20–30
38. Eisen, J.S., Myers, P.Z. and Westerfield, M. (1986) Nature (London) **320**, 269–271
39. Holt, C.E. (1989) J. Neurosci. **9**, 3123–3145
40. Bonhoeffer, F. and Huf, J. (1980) Nature (London) **288**, 162–164
41. Haydon, P.G., McCobb, D.P. and Kater, S.B. (1984) Science **226**, 561–564
42. Mattson, M.P., Taylor-Hunter, A. and Kater, S.B. (1988) J. Neurosci. **8**, 1704–1711
43. Mattson, M.P., Dou, P. and Kater, S.B. (1988) J. Neurosci. **8**, 2087–2100
44. Mattson, M.P. and Kater, S.B. (1989) Brain Res. **490**, 110–125
45. Davenport, R.W. and McCaig, C.D. (1993) J. Neurobiol. **24**, 89–100
46. Eisen, J.S. (1994) Annu. Rev. Neurosci. **17**, 1–30
47. Davenport, R.W., Löschinger, J., Huf, J., Jung, J. and Bonhoeffer, F. (1994) Soc. Neurosci. Abstr. **20**, 1065
48. Zheng, J., Zheng, Z. and Poo, M.-M. (1994) J. Cell Biol. **127**, 1693–1701
49. Grynkiewicz, G., Poenie, M. and Tsien, R. (1985) J. Biol. Chem. **260**, 3440–3450
50. Tsien, R.Y. (1988) Trends Neurosci. **11**, 419–424
51. Silver, R.A., Lamb, A.G. and Bolsover, S.R. (1989) J. Neurosci. **9**, 4007–4020
52. Connor, J.A. (1986) Proc. Natl. Acad. Sci. U.S.A. **83**, 6179–6183
53. Cohan, C.S., Haydon, P.G. and Kater, S.B. (1985) J. Neurosci. Res. **13**, 285–300
54. Rehder, V. and Kater, S. (1992) J. Neurosci. **12**, 3175–3186
55. Mattson, M.P., Guthrie, P.B. and Kater, S.B. (1988) J. Neurosci. Res. **21**, 447–464
56. Lankford, K.L. and Letourneau, P.C. (1989) J. Cell Biol. **109**, 1229–1243
57. Lankford, K.L. and Letourneau, P.C. (1991) Cell Motil. Cytoskel. **20**, 7–29
58. Cohan, C.S., Connor, J.A. and Kater, S.B. (1987) J. Neurosci. **7**, 3588–3599

59. Haydon, P.G. and Kater, S.B. (1987) in Growth and Plasticity of Neural Connections (Winlow, W. and McCrohan, C., eds.), pp. 8–26, Manchester University Press, Manchester
60. Mattson, M.P., Lee, R.E., Adams, M.E., Guthrie, P.B. and Kater, S.B. (1988) Neuron **1**, 865–876
61. McCobb, D.P. and Kater, S.B. (1988) Dev. Biol. **130**, 599–609
62. Cohan, C.S. (1992) J. Neurobiol. **23**, 983–996
63. Fields, R., Guthrie, P., Russell, J. et al. (1993) J. Neurobiol. **24**, 1080–1089
64. Fields, R. and Nelson, P. (1994) J. Neurobiol. **25**, 281–293
65. Goodman, C.S. and Shatz, C.J. (1993) Cell **72**/Neuron **10** (Suppl.), 77–98
66. Shatz, C.J. (1990) Neuron **5**, 745–756
67. Ivins, J., Raper, J. and Pittman, R. (1991) J. Neurosci. **11**, 1597–1608
68. Kater, S.B., Mattson, M.P., Cohan, C.S. and Conner, J. (1988) Trends Neurosci. **11**, 315–321
69. Kater, S. and Mills, L. (1991) J. Neurosci. **11**, 891–899
70. Nishi, R. and Berg, D.K. (1981) Dev. Biol. **87**, 301–307
71. Anglister, L., Farber, I., Shahar, A. and Grinvald, A. (1982) Dev. Biol. **94**, 351–365
72. Cohan, C.S. (1990) J. Neurobiol. **21**, 400–413
73. Hantaz-Ambroise, D. and Trautmann, A. (1989) Int. J. Dev. Neurosci. **7**, 591–602
74. Robson, S.J. and Burgoyne, R.D. (1989) Neurosci. Lett. **104**, 110–114
75. Fields, R.D., Neale, E.A. and Nelson, P.G. (1990) J. Neurosci. **10**, 2950–2964
76. Fields, R. and Nelson, P. (1994) Physiol. Chem. Phys. Med. NMR **26**, 27–53
77. Al-Mohanna, F., Cave, J. and Bolsover, S.R. (1992) Dev. Brain Res. **70**, 287–290
78. Tsien, R.Y. (1980) Biochem. J. **19**, 2396–2404
79. Silver, R.A., Lamb, A.G. and Bolsover, S.R. (1990) Nature (London) **343**, 751–754
80. Bedlack, R.S., Jr., Wei, M.-D. and Loew, L.M. (1992) Neuron **9**, 398–404
81. Davenport, R.W. and Kater, S.B. (1992) Neuron **9**, 405–416
82. Goldberg, D.J. (1988) J. Neurosci. **8**, 2596–2605
83. McCaig, C.D. (1986) J. Physiol. **375**, 55–69
84. McCaig, C.D. (1989) J. Cell Sci. **93**, 723–730
85. Robinson, K.R. (1985) J. Cell Biol. **101**, 2023–2027
86. McCaig, C.D. and Rajnicek, A.M. (1991) Exp. Physiol. **76**, 473–494
87. Wessells, N.K. and Nuttall, R.P. (1978) Exp. Cell Res. **115**, 111–122
88. Mills, L.R., Murrain, M., Guthrie, P.B. and Kater, S.B. (1988) Soc. Neurosci. Abstr. **14**, 583
89. O'Connor, T.P., Duerr, J.S. and Bentley, D. (1990) J. Neurosci. **10**, 3935–3946
90. Nakai, J. and Kawasaki, Y. (1959) Zeitschrift fur Zellforschung **51**, 108–122
91. Heidemann, S.R., Lamoureux, P. and Buxbaum, R.E. (1990) J. Cell Biol. **111**, 1949–1957
92. Bray, D. (1973) Nature (London) **244**, 93–96
93. Bray, D. (1979) J. Cell Sci. **37**, 391–410
94. Hammarback, J.A. and Letourneau, P.C. (1986) Dev. Biol. **117**, 655–662
95. Gundersen, R.W. (1988) J. Neurosci. Res. **21**, 298–306
96. Letourneau, P.C. and Shattuck, T.A. (1989) Development **105**, 505–519
97. Davenport, R.W., Dou, P., Rehder, V. and Kater, S. (1993) Nature (London) **361**, 721–724
98. Lipscombe, D., Madison, D.V., Poenie, M. et al. (1988) Proc. Natl. Acad. Sci. U.S.A. **85**, 2398–2402
99. Haydon, P.G. and Man-Son-Hing, H. (1988) Neuron **1**, 919–927
100. Mills, L.R. and Kater, S.B. (1990) Neuron **4**, 149–163

6

Mechanisms of growth cone guidance by proteoglycans

Diane M. Snow*‡, Paul B. Atkinson†, Tim D. Hassinger†, S.B. Kater† and Paul C. Letourneau*

*Department of Cell Biology and Neuroanatomy,
The University of Minnesota, Minneapolis, MN 55455, U.S.A.
and †Department of Anatomy and Neurobiology,
The University of Utah, Salt Lake City, UT 84132, U.S.A.

Introduction

During development of the nervous system, the sensory and locomotory apparatus of the elongating neurite, the growth cone, must navigate to its appropriate target by means of continuous interactions with the extracellular matrix (ECM), neighbouring cell surfaces and secreted factors. A vast number of molecules have been identified that are expressed within the neuron's extracellular milieu, and are suggested to have specific effects on neuronal behaviour, many of which occur via complex signalling cascades [1–7]. Although much progress has been made, elucidation of the interactions of these molecules with other components of the nervous system, and identification of the precise role of each of these molecules remains a difficult and major goal of developmental neurobiology.

Proteoglycans are one type of molecule expressed during neural development

One class of molecule that has been increasingly implicated in the development and maintenance of the nervous system are the proteoglycans (PGs). PGs are a class of glycoprotein that consist of a protein core decorated with one or more types of glycosaminoglycan (GAG) [8,9]. These complex molecules (i) are expressed throughout the developing and mature nervous system, as well as in the injured adult nervous system [10–14], (ii) are found in all cellular locations from the nucleus to the ECM [15], and (iii) interact with a wide variety of other molecules [16–20]. Although PGs have traditionally been considered a 'class' of molecule, this distinction is based solely on the fact that their general structure consists of a core protein and carbohydrate side chains. In actuality, PGs differ greatly, not only in their number and type of carbohydrate chains, but also in the composition of their

‡*To whom correspondence should be addressed.*

core protein [9,21]. Cloning and sequencing of numerous PG core proteins do not show significant sequence similarities between different PG cores [in fact, PG core proteins share much similarity with other molecules such as cell adhesion molecules (CAMs) and growth factors], making it awkward to consider these diverse species of molecules as a class or family [22–25]. Continued cloning and sequencing of PG core proteins will dictate how they should be grouped on a genetic level.

Although PGs have been well characterized in such systems as cartilage and connective tissue [26–28], their role in the nervous system has been a focus of more recent interest [29]. The investigation of PGs in the nervous system is currently shedding new light on their structure, interactions and functions. Reports show that PGs influence developing neurons in numerous and contrasting ways. Some reports suggest that PGs may positively influence neurite outgrowth [30–42]. Many such studies suggest a growth-promoting role for PGs, because they have been demonstrated to be present within axon pathways. However, the precise interactions of these PGs with other components within the axon pathways have not been well characterized and may play a major role in their function. Other studies indicate that certain PGs inhibit neurite outgrowth *in vivo* and *in vitro* [43–60], as well as inhibit cell migration [61,62], and cell attachment and spreading [63–65].

The above studies indicate a multifunctional capacity for PGs in the developing nervous system, and elsewhere. This multifunctionality may be attributable to a number of factors, such as (1) variability in the abundance or distribution of PGs at a given location; (2) whether a cell interacts with the carbohydrate or protein component of the PG, or with both; (3) the interactions of the PG with other molecules; or (4) it may be due to functions of different molecules that associate with PGs.

Possible mechanisms for PG regulation of growth cone behaviour

Although the core proteins, and/or GAG composition of many PGs have been identified, isolated, characterized, and/or cloned [22,23,66–70], the distinct mechanisms by which these molecules act remain unclear. Speculation as to how PGs may influence neurite outgrowth include many mechanisms. PGs may block adhesive ligands specifically or via steric hindrance, e.g. by virtue of their water-binding capacity or extensive carbohydrate chains. They may act by binding to and presenting growth factors at appropriate places and times during development, or by binding to other molecules resulting in conformational changes [70a]. PGs may act by creating extracellular spaces, by presenting a high density of negative charge, or perhaps through specific receptors [58,71]. Obviously, PGs may function through a combination of these mechanisms as well.

Cytoplasmic calcium can mediate changes in growth cone behaviour

A large body of literature shows that one means by which growth cones undergo changes in morphology and direction is through the second messenger cytoplasmic Ca^{2+} level [72–74]. Changes in cytoplasmic calcium ion concentration ($[Ca^{2+}]_i$) can affect both lamellipodial and filopodial extension, and growth cone turning [75–78]. Important to the present study, the elevation of $[Ca^{2+}]_i$ has been shown to inhibit neurite elongation in neuroblastoma cells [79], and elevated $[Ca^{2+}]_i$ leads to the breakdown of actin filaments [80]. Studies showing a role for a variety of signalling pathways in the facilitation of growth cone behaviours are ever increasing, and provide a rich area of study to determine the mechanisms by which growth cones adhere to their substrata, elongate and select their appropriate targets [1,81–83]. Whether specific signalling pathways are involved in PG regulation of growth cone behaviour is the focus of this and other investigations.

Does chondroitin sulphate PG (CSPG) alter cytoplasmic calcium?

In light of the above findings, we chose to examine whether changes in $[Ca^{2+}]_i$ might act as a signal to neurons that contact PGs. Specifically, we used the calcium indicator fura-2/acetoxymethyl (AM) ester to monitor $[Ca^{2+}]_i$ in chick dorsal root ganglion (DRG) neuronal growth cones during contact with a specific PG, CSPG: (1) to determine whether there is a change in $[Ca^{2+}]_i$ in neurons that contact CSPG; and (2) to determine whether changes in $[Ca^{2+}]_i$ are necessary for inhibition of growth cone migration.

The following report summarizes studies showing that sensory neurons respond to CSPG contact with a transient rise in $[Ca^{2+}]_i$. Thus, changes in growth cone $[Ca^{2+}]_i$ could provide a signalling pathway through which sensory neurons respond to PGs *in vivo*. The effect of CSPG contact is concentration-dependent and requires the carbohydrate moiety of CSPG, but not the core protein. Further, the addition of soluble CSPG to the medium does not elevate $[Ca^{2+}]_i$. Treatment with reagents that block plasma membrane calcium channels, or that perturb intracellular Ca^{2+} stores, indicate that extracellular Ca^{2+} is the major source of the $[Ca^{2+}]_i$ elevation, and that Ca^{2+} entry occurs through non-voltage-gated calcium channels. In preliminary studies, general Ca^{2+}-channel blockers did not abolish growth cone avoidance of surface-bound CSPG. We conclude therefore that (1) sensory neurons, in response to CSPG contact, elevate $[Ca^{2+}]_i$ to levels that can modify cytoskeletal mechanisms of growth cone migration, and (2) avoidance of substratum-bound CSPG may not be dependent upon elevated $[Ca^{2+}]_i$.

CSPGs can signal via calcium ions in DRG cell bodies and growth cones

The primary question asked in these studies was whether a PG could stimulate changes in $[Ca^{2+}]_i$, i.e. act as a signalling molecule, and thereby potentially regulate growth cone behaviour. Since numerous reports have shown that CSPGs can act as inhibitors of growth cone elongation [10,43,45,47,51,56,59], our interests focused on whether signalling via cytoplasmic calcium might regulate growth cone inhibition by PGs. The basic paradigms used are described below.

Basic techniques

The techniques employed to assay calcium changes in DRG neurons in response to CSPG were as follows. For the majority of experiments, glass coverslips (24 mm × 40 mm) were acid-washed overnight, rinsed with distilled water and oven-dried at 100 °C. The coverslips were mounted over holes drilled in 60-mm-diam. culture dishes (Falcon), containing a mixture of petroleum, bee's wax and paraffin (1:1:1, by volume). All dishes were UV-treated for 30 min and subsequently treated as sterile. The coverslips were coated with 25 μg/ml filter-sterilized laminin, or fibronectin, for at least 3 h at room temperature. Laminin isolated from Engelbreth-Holm-Swarm sarcoma cells (EHS laminin) was supplied by S. Palm (Department of Laboratory Medicine and Pathology, University of Minnesota), and fibronectin was supplied by J. McCarthy (Department of Laboratory Medicine and Pathology, University of Minnesota). Prior to use, laminin- or fibronectin-coated coverslips were repeatedly rinsed in PBS followed by rinses with sterile distilled water, then either covered with medium and stored until use, or air-dried and adsorbed with CSPG (as described below). Results of experiments using a laminin-coated substratum were indistinguishable from those using a fibronectin-coated substratum in these assays.

For the majority of experiments, lumbar DRG neurons were dissected from embryonic day (E) 8–10 chickens into F12 culture medium supplemented with 10% (v/v) fetal-calf serum (F12HS10). DRG neurons were cleaned of surrounding tissue and dissociated to single cells by treatment with 0.25% crude bovine trypsin in 0.1 M calcium/magnesium-free phosphate-buffered saline (CMF-PBS; pH 7.89) for 18 min at 37 °C, resuspended and triturated in 10 mM Hepes-buffered F14 medium supplemented with L-glutamate (2 mM), sodium selenite (5 ng/ml), sodium pyruvate (200 μg/ml), phosphocreatine (5 mM), progesterone (20 nM), insulin (5 μg/ml), transferrin (100 μg/ml) and nerve growth factor (NGF; 50 ng/ml), and an antibiotic/antimycotic solution (penicillin/streptomycin/fungizone; PSF); or, for some experiments, in calcium-free medium prepared by adding essential amino acids, vitamins, $MgCl_2$, $MgSO_4$ and $NaHCO_3$ to Calcium/Magnesium-free Hanks Balanced Salt Solution (CMF-HBSS) with 1 mM EGTA, containing calcium-free NGF (30 ng/ml). In some cases, E6–7 chick retinal ganglion cells (RGCs) were used and were dissected as described in [51]. In addition, lanthanum chloride (1 mM), a general calcium-channel blocker, was prepared in medium lacking carbonate, sulphate and phosphate to prevent precipitation of polyvalent cations. Approx. 2×10^4 cells were plated in 500 μl of medium. Cells were incubated in an ambient chamber at 40 °C for at least 6 h.

The first paradigm used to examine DRG neuron contact with CSPG involved the use of polystyrene beads, to which CSPG was adsorbed, and is referred to herein as the 'CSPG bead technique' (Fig. 1A). The 'CSPG stripe technique' (Fig. 1B) will be described later. The various forms of CSPG adsorbed were: (1) purified bovine nasal cartilage proteoglycan (chondroitin sulphate/keratan

---

Fig. 1   **Two paradigms were used to examine DRG contact with bound CSPG**

(A) CSPG Bead Technique

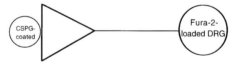

(B) CSPG Stripe Technique

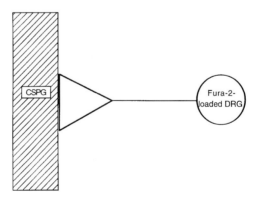

(**A**) The CSPG bead technique involves the use of the Laser Tweezer 2000 (Cell Robotics, Inc.) laser trap to capture and move a CSPG-coated bead to elongating DRG neurons. In some cases, beads were captured with a micropipette and moved with a hydraulic micromanipulator. (**B**) The CSPG stripe technique is a guidance assay where an elongating DRG neuron encounters a stripe of substratum-bound CSPG. In both cases, DRG neurons were loaded with the calcium indicator fura-2/AM, and calcium changes within the growth cone, axon and/or cell body were monitored using fluorescence imaging techniques (for details of methods see the text).

---

sulphate-PG), which contains chondroitin-0-sulphate, chondroitin-6-sulphate, a small contribution from chondroitin-4-sulphate and keratan sulphate (KS), and O- and N-linked oligosaccharides (aggrecan-like [84]), provided by L. Culp (Case Western Reserve University, Cleveland, OH) (for characterization see [84a]; (2) chick limb bud CSPG, provided by D. Carrino and A. Caplan, CWRU (for characterization see [84b]; and (3) bovine aorta CSPG (Collaborative Biomedical Products, Cat no. 40252; versican-like [85]). The polystyrene microspheres (Polysciences, Inc.; 3–10 μm; 300 μl volume) were washed three times in 0.1 M PBS. A 400 μl solution of beads was incubated overnight at 4 °C in 1 mg/ml CSPG, resulting in the non-covalent attachment of CSPG to the bead surface. CSPG-coated beads were stored at 4 °C in 50 μl aliquots until required. Immuno-cytochemistry was used [47] to determine that the CSPG bound to the beads. Binding of CSPG to beads was quantified using 35S-chick limb bud CSPG (generously provided by D. Carrino and A. Caplan), and indicated that 5–10% of the CSPG in the incubation solution bound to the beads. This corresponds to an approximate surface density of CSPG binding to beads of 0.0005 pg/μm^2, which is in the order of the surface density of CSPG in the stripe assay (0.003 pg/μm^2) [47,54]. In later experiments, CSPG was covalently bound to carboxylated beads using the Polysciences, Inc. carbodiimide kit (19539), and gave similar results.

CSPG-coated beads were touched to DRG neuron cell bodies by one of two methods. Initially, beads were selected with a micropipette then moved by a hydraulic micromanipulator (Narishige MO-103) to the elongating growth cone. Throughout the remainder of the experiments, a more sophisticated means of bead manipulation employed the Laser Tweezers 2000 (Cell Robotics, Inc., Albuquerque, NM, U.S.A.), a laser trap used to capture beads and move them from one location in a culture dish to another. This technique provided much finer precision and accuracy of bead mobilization and contact.

Prior to taking calcium measurements, DRG neurons were incubated in medium containing 2–5 μM fura-2/AM (Molecular Probes, Inc.) for 30–40 min. Cells were washed several times and incubated 30 min at 37 °C in medium alone to allow for de-esterification of fura-2 before calcium levels were measured. Images were viewed with either a × 40, × 63 or × 100 fluor/phase-contrast oil-immersion objective and appropriate neutral density filters, as well as a Uniblitz shutter (Victor Associates, Inc.) to ensure minimum UV exposure to the cells. For most experiments, fura-2 fluorescence imaging was done using an intensified (change - coupled device; CCD) camera (PaulTek) and the Image-1 fluorescence analysis system (Universal Imaging, Inc., West Chester, PA, U.S.A.). For some experiments, phase-contrast and fluorescence emission images were acquired using a (CCD) camera (Photometric, Inc.), and images of fluorescence emission in response to 350 and 380 nm excitation were compared on a MacIntosh IIfx computer. Calcium measurements were taken prior to CSPG contact, then beads were touched to the cells and pulled away while calcium measurements were taken at the moment of contact. Fig. 2(A) is a phase-contrast photograph demonstrating multiple CSPG-coated bead contact with a DRG cell body. Calcium was then measured at approx. 30 s to 1 min intervals following CSPG contact. Phase-contrast images were recorded simultaneously with calcium measurements in many cases (Fig. 2A). Images were captured every 10–30 s. Quantitative calcium values were calculated according to the calcium equation given by Grynkiewicz *et al.* [86], $K_d = 224$ nM, in a system calibrated *in vitro* or *in situ* (in living cells).

Fig. 2. **DRG cell bodies respond to CSPG contact with an elevation of $[Ca^{2+}]_i$**

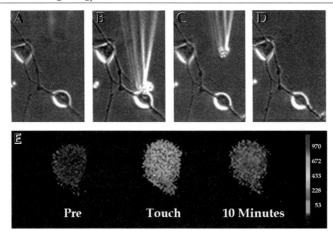

(*A–D*) Phase-contrast microscopy demonstrating the method of presentation of CSPG-coated beads to DRG cell bodies. (*A*) Cell bodies on laminin prior to CSPG contact, (*B*) a cell body touched with CSPG-coated beads, (*C*) a cell body immediately after, and (*D*) 5–20 min after CSPG-coated beads are pulled away. (*E*) Fluorescence microscopy of the calcium indicator fura-2/AM to image $[Ca^{2+}]_i$ in DRG cell bodies that contact CSPG. A DRG cell body on laminin (Pre) elevates $[Ca^{2+}]_i$ in response to contact with CSPG-coated beads (Touch). CSPG contact induced a rise in $[Ca^{2+}]_i$ in 86.4% of DRG cell bodies with the mean $\Delta[Ca^{2+}]_i = 517 \pm 115$ nM; n = 24; P <0.0001. Within 10 min, the DRG cell body recovered toward baseline. DRG cell bodies are approx. 10 μm. [Modification of (*E*) reproduced from [60] with permission from Academic Press, Inc.]

A general criterion was established in order to determine the specificity of CSPG-induced calcium increases. Since simply touching neurons can result in rises of the order of 10–20 nM, dependent upon cell type, we set a criterion level for positively scoring an induced increase in intracellular calcium at a minimum of 50 nM. In all cases, CSPG alone was nearly an order of magnitude higher than this criterion, and in most cases, controls prove to be nearly an order of magnitude lower than this criterion.

In controls, cell bodies and growth cones were touched with: (1) a glass micropipette alone (mean $\Delta[Ca^{2+}]_i = 25.5 \pm 1.5$ nM; $n = 6$; $P > 0.1$), or (2) beads

Fig. 3.   **Fluorescence microscopy with the fluorescent calcium indicator fura-2/AM to image [Ca²⁺]ᵢ in DRG growth cones**

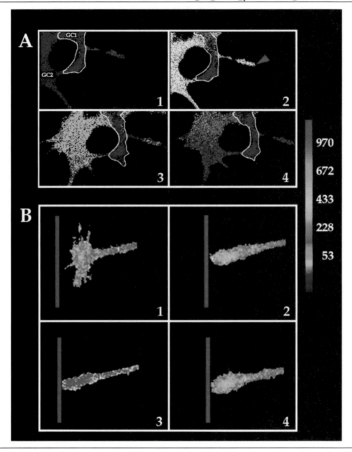

(**A**) Two overlying growth cones prior to CSPG contact (panel 1). A filopodium of growth cone 2 (GC2) is touched with CSPG beads (arrowhead in panel 2) and simultaneously elevates [Ca²⁺]ᵢ. Within minutes, calcium levels recover toward baseline (panel 3), and have returned to baseline by 3 min (panel 4). CSPG contact with GC2 does not induce elevation of [Ca²⁺]ᵢ in the overlying growth cone (GC1). (**B**) An elongating DRG growth cone (panel 1) first contacts a CSPG-adsorbed stripe (panel 2) via one or more filopodia. Although some filopodia retract upon CSPG contact, many filopodia still remain active at the CSPG border; CSPG does not routinely cause growth cone collapse (see [51]). When the growth cone makes greater contact with CSPG, i.e. a greater amount of surface area of the growth cone comes into contact with CSPG (panel 3), [Ca²⁺]ᵢ is elevated to the micromolar range. Elevation of [Ca²⁺]ᵢ occurred in 83.3% (n = 22) of the growth cones that contacted CSPG-adsorbed stripes (mean Δ[Ca²⁺]ᵢ = 594 ± 1 02 nM; P <0.0001). This growth cone exhibits a slow recovery toward baseline (panel 4). See text for analysis of recovery phases, and Fig. 4(D) (closed circles) for a graph of the time course for (**B**). Reproduced from [60] with permission from Academic Press, Inc.

incubated in PBS {PBS is the diluent for CSPG; mean $\Delta[Ca^{2+}]_i = 78 \pm 2.5$ nM; $n = 4$; $P > 0.05$; multiple beads (three or more) were used in all bead controls}. Furthermore, to rule out bias we performed a blind study where the experimenter did not know whether the beads were CSPG-coated or incubated in PBS only (mean $\Delta[Ca^{2+}]_i = 1500$ nM with CSPG-coated beads; mean $\Delta[Ca^{2+}]_i = 82$ nM with PBS-coated beads; $n = 3$; $P > 0.1$). In all control experiments, treatments did not result in elevation of $[Ca^{2+}]_i$ in DRG neurons. A further control to test specificity used enzyme degradation of the carbohydrate moiety, and is described below.

In addition to taking calcium measurements, DRG neurons were observed by light microscopy using an inverted microscope (IM 35; Carl Zeiss, Inc., Thornwood, NY, U.S.A.) warmed to 40 °C with an air curtain incubator (ASI 400; Carl Zeiss, Inc.). Growth cone behaviour was monitored by phase-contrast optics using a Newvicon video camera (NC-65; Dage-MTI, Inc., Michigan City, IN, U.S.A.), and image analysis software (Image 1; Universal Imaging, Inc., West Chester, PA, U.S.A.), which was run on a 486/33 computer system (Gateway 2000, North Sioux City, SD, U.S.A.). Images were viewed with a monitor (Trinitron; Sony Corp. of America, New York, NY, U.S.A.) and recorded every 10–60 s with an optical disc video recorder (TQ-2026F; Panasonic Industrial Comp., Secaucus, NJ, U.S.A.).

Results of the CSPG bead technique

The majority of DRG cell bodies responded to CSPG contact with a transient rise in $[Ca^{2+}]_i$ (mean $\Delta[Ca^{2+}]_i$ above resting level was 554 ± 109 nM; $P < 0.0001$) (Fig. 2B). $[Ca^{2+}]_i$ became elevated with CSPG contact, then returned to baseline levels following removal of the CSPG-coated beads. This important result demonstrated that contact with CSPG can indeed trigger the elevation of $[Ca^{2+}]_i$ in DRG neurons, and therefore could potentially act as a signalling molecule in these cells.

There was no obvious difference in the response to CSPG contact with cell bodies compared with contacts with growth cones, since contact of DRG growth cones with CSPG resulted in rises of the same magnitude as for cell bodies: mean $\Delta[Ca^{2+}]_i = 548 \pm 66$ nM; $n = 22$; $P < 0.0001$ (Fig. 3A). (Fig. 3B shows the elevation of $[Ca^{2+}]_i$ in growth cones that contact CSPG in a guidance assay, not bound to beads, and will be described later.) Further characterization of the effects of CSPG-coated beads on $[Ca^{2+}]_i$ combines results of stimulation to cell bodies and growth cones of DRG neurons.

The response of RGC growth cones to CSPG-coated beads was quite different from that of DRG growth cones. RGC growth cones showed a graded response, with highest elevation of $[Ca^{2+}]_i$ near the site of bead contact and diminished elevation of $[Ca^{2+}]_i$ with increasing distance from the contact site (D.M. Snow, unpublished work). This is in contrast to the elevation of $[Ca^{2+}]_i$ that fills the entire growth cone and sometimes extends into the axon for DRG neurons. Although the implications of this difference between the DRG and RGC calcium response for neurite outgrowth are unclear, it may relate to the differences we see in the behaviour of these two neuronal types both for the CSPG single stripe assay [47], and the CSPG step gradient assay [54]. These differences will be examined in greater detail in future studies.

Possible roles for the elevation of $[Ca^{2+}]_i$ in sensory neurons

There are many cellular functions that could be directly affected by a rise in sensory neuron growth cone $[Ca^{2+}]_i$. Our particular interests have led us to focus on the relationship between elevated $[Ca^{2+}]_i$ and growth cone navigation. Previous work shows that the CSPG-induced elevation of $[Ca^{2+}]_i$ which we observed in DRG growth cones is within the range of calcium increases that significantly alter the structural components necessary for growth cone behaviour [75,77,80,87–89]. Specifically, elevation of cytoplasmic calcium to the micromolar range in growth cones can destabilize peripheral actin. On the basis of these results, we propose that calcium might fulfil sufficient conditions to act as an intermediate in PG-induced inhibition of growth cone migration, possibly by promoting reorganization of the cytoskeleton for stopping and turning at a CSPG border.

Elevation of $[Ca^{2+}]_i$ in response to contact with CSPG may signal growth cone cytoskeletal reorganization. First, contact with CSPG may have direct effects on the cytoskeleton that increase the dynamic motility of the growth cone. In the leading margin of the growth cone, actin filaments are broken down in response to an elevation of $[Ca^{2+}]_i$ [80,88,90]. Since filopodia continuously sample a CSPG-adsorbed substratum but do not collapse and retract, such an action on actin filaments may regulate the behaviour of filopodia to promote growth cone turning.

Actin filaments in growth cones may be cross-linked by actin-binding proteins such as α-actinin and filamin [91], which are also affected by changes in $[Ca^{2+}]_i$ [92]. When $[Ca^{2+}]_i$ is elevated, α-actinin dissociates from actin filaments, which could break down the filament network and destabilize filopodia and lamellipodia [93–96]. Gelsolin, another actin-binding protein, is activated by elevated $[Ca^{2+}]_i$ to sever actin filaments, and cap the barbed, or growing, end of actin filaments [97]. Immunocytochemical evidence indicates that gelsolin is present in the leading margin of DRG growth cones (P.C. Letourneau, unpublished work). Thus, the effects of elevated $[Ca^{2+}]_i$ on actin-binding proteins could lead to a rapid turnover of filopodia and lamellipodia, and thereby facilitate redirection of the growth cone.

The role of actin in growth cone behaviour in response to changes in $[Ca^{2+}]_i$ probably involves other cytoskeletal components, such as microtubules. For example, in experiments where actin localization is continuously monitored during growth cone contact with CSPG, we observed that as actin-based filopodia touched CSPG and withdrew dynamic microtubules extended into the central region of the growth cone (overlapping with actin filaments at the base of filopodia), while stable microtubules were arrested at the border between CSPG and laminin [98]. Thus, changes in filopodia $[Ca^{2+}]_i$ may play a role in signalling associated cytoskeletal components to direct changes in DRG migration.

In addition to direct effects that stimulate cytoskeletal reorganization, elevation of $[Ca^{2+}]_i$ in growth cones that contact CSPG may affect the motile apparatus of the growth cone by triggering other multicomponent signalling cascades. Prolonged responses within growth cones may be activated in this manner. Neurons contain other major cytosolic targets for Ca^{2+}, e.g. protein kinase C (PKC) [3,99], calpain [100] and calmodulin (CaM) [101,102]. When activated, PKC can phosphorylate and thereby regulate membrane proteins. Similarly, the Ca^{2+}-dependent protease calpain, when activated, can proteolyse membrane proteins, and has been shown to regulate cytoskeletal dynamics in neurons [100]. CaM is a Ca^{2+}-binding regulatory protein that is progressively activated as intracellular calcium is raised above resting levels, with maximum binding of calcium at the micromolar range [101]. An important function of activated CaM is to bind to Ca^{2+}/CaM-dependent protein kinases, which in turn phosphorylate a variety of proteins. In transgenic *Drosophila* mutants where CaM could not bind to target proteins or to itself, growth cone guidance errors resulted, suggesting a role for CaM in growth cone navigation [103].

The overall result of the elevation of $[Ca^{2+}]_i$ in growth cones is dependent upon the spatial organization of target proteins and their affinities for Ca^{2+}, as well as the location of downstream components of signalling cascades. Furthermore, the state of actin is not just predicted from the moment to moment $[Ca^{2+}]$, but from the temporal course of $[Ca^{2+}]$ change, the spatial distribution of calcium in the growth cone, and the involvement of other signalling pathways that are simultaneously activated [92]. Thus, the elevation of $[Ca^{2+}]_i$ elicited by CSPG contact may signal multiple changes simultaneously via actin-binding proteins, PKC, calpain, CaM and/or related proteins that facilitate turning at a boundary with CSPG.

Another interpretation of our data is that CSPG may inhibit growth cone advance by a mechanism unrelated to changes in $[Ca^{2+}]_i$, and that the increase in $[Ca^{2+}]_i$ is acting as a signal for a function not yet described. Given that PGs could play a role in signalling events such as synaptogenesis [104], and in developmental plasticity [105], this possibility is plausible, and will require further testing.

Characteristics of the CSPG-induced elevation of $[Ca^{2+}]_i$

CSPG-induced elevation of $[Ca^{2+}]_i$ requires the carbohydrate component of CSPG

The inhibitory response of DRG and RGC growth cone migration to CSPG *in vitro* is dependent upon the carbohydrate component of CSPG [45,47,51]. Therefore, CSPG-coated beads were treated with a mixture of chondroitinase ABC, and keratanase (endo-β-galactosidase) purchased from Miles Scientific (Cat. nos. 32-030 and 32-032, respectively) to remove all carbohydrate moieties from CSPG. They were then touched to DRG neurons to determine if the carbohydrate portion of CSPG is also necessary for elevation of $[Ca^{2+}]_i$ (Fig. 4A). The removal of

Fig. 4.   **Characteristics of the CSPG response in DRG neurons**

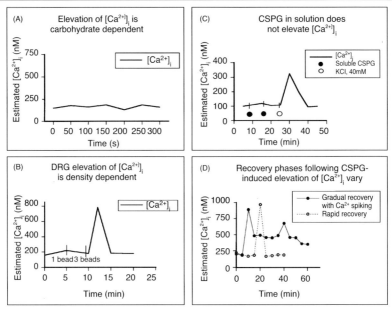

(A) CSPG-induced elevation of $[Ca^{2+}]_i$ requires the carbohydrate component of CSPG. DRG growth cones and cell bodies were touched with CSPG-coated beads treated with enzymes that completely remove the carbohydrate moieties of CSPG. $[Ca^{2+}]_i$ is not elevated by contact with enzyme-treated beads (mean $\Delta[Ca^{2+}]_i = 18.5 \pm 5.5$ nM; n = 4; P > 0.5). DRG growth cone is approx. 10 μm long. (B) CSPG-induced elevation of $[Ca^{2+}]_i$ is dependent on a threshold number of beads contacting the growth cone. Contact between growth cones and a single 3-μm-diam. CSPG-coated bead does not usually result in the elevation of $[Ca^{2+}]_i$, while contact with a cluster of 3-μm-diam. CSPG-coated beads (three or more) results in a large increase in $[Ca^{2+}]_i$. (C) CSPG in solution does not induce a rise in $[Ca^{2+}]_i$. The addition of 1–1000 μg/ml CSPG to the medium (black circles) in cultures of DRG neurons does not inhibit normal elongation, nor does $[Ca^{2+}]_i$ become elevated. The addition of 40 mM KCl (open circle) demonstrates that the cells are competent to increase $[Ca^{2+}]_i$. (D) CSPG-induced elevation of $[Ca^{2+}]_i$ results in two types of recovery phases. Elevation of $[Ca^{2+}]_i$ occurs immediately upon contact with CSPG. However, the recovery phase for these cells is variable. In some cases, recovery is fast (<5 min; mean = 3 min 43 s; n = 8) where $[Ca^{2+}]_i$ returns to the original baseline quickly and is maintained at that level (36.8%; n = 19) (open circles). In other cases, recovery is slow (>10 min; mean = 23 min 20 s; n = 3), and $[Ca^{2+}]_i$ returns to a baseline that is higher than the original (63.2%, n = 19) (closed circles). Intermittent Ca^{2+} transients often occurred during the slow recovery phase. (Note: The trace with closed circles is a graphic representation of the growth cone shown in Fig. 3B using the CSPG stripe technique).

carbohydrate from CSPG-beads resulted in a change from a mean $\Delta[Ca^{2+}]_i$ for untreated CSPG-coated beads of 517 ± 115 nM, to a mean $\Delta[Ca^{2+}]_i$ for enzyme-treated beads of 18.5 ± 5.5 nM ($n = 4$; $P > 0.5$). Thus, as for growth cone guidance, the carbohydrate component of CSPG is required for the elevation of $[Ca^{2+}]_i$. The mechanism by which chondroitin sulphate carbohydrate induces elevation of $[Ca^{2+}]_i$ is unknown, but an important factor may be the sulphation pattern and/or resultant negative charge distribution of the carbohydrate chains. This prospect will be examined in future studies.

CSPG-induced elevation of $[Ca^{2+}]_i$ is related to the number of CSPG-coated beads, i.e. the density of CSPG

Inhibition of growth cone migration by a CSPG-adsorbed stripe is dependent on the density of CSPG binding [47,54]. In the CSPG bead assays, contact with one or two 3-μm-diam. beads was not sufficient to elicit a $[Ca^{2+}]_i$ rise in most cases, while three or more 3-μm-diam. CSPG-coated beads induced elevation of $[Ca^{2+}]_i$ to commonly obtained levels (Fig. 4B). The approximate surface density of CSPG on the beads was 0.0005 pg/μm^2, which is in the order of the surface density of CSPG in the stripe assay (0.003 pg/μm^2). Because of the curved surfaces of the beads, a single filopodium may contact only a limited region of the CSPG-containing surface of a bead at one time, whereas it could interact with more CSPG on a flat substratum. However, the capability of a filopodium to raise and lower from the substratum may indeed allow it to interact to a large extent with the CSPG-coated bead surface [106]. The requirement of three or more beads to reach a threshold to elevate $[Ca^{2+}]_i$ suggests that the stimulus to induce elevation of $[Ca^{2+}]_i$ in DRG neurons requires a sufficient surface area of growth cone contact by CSPG-coated beads, or possibly, that multiple discrete stimulus sites are required. *In vivo*, several filopodia or the lamellipodia of a growth cone may contact CSPG at the same time. (See the requirement of veil contact for elevation of $[Ca^{2+}]_i$ in Fig. 3B).

CSPG in solution does not elevate $[Ca^{2+}]_i$ in DRG neurons

Studies from our laboratory indicate that CSPG added in solution to neurons growing on laminin does not inhibit neurite outgrowth. However, CSPG added in solution to cells growing on fibronectin does affect some aspects of neuronal behaviour [70a]. Furthermore, CSPG added to the culture medium at the time of seeding DRG neurons reduces adhesion to an extent proportional to the CSPG concentration [108]. Similarly, other studies have shown that soluble PG, or their carbohydrate side chains (GAGs) in solution, can affect growth cone behaviour [34,36,49,50]. Consistent with our observation that soluble CSPG added to DRG growing on laminin did not inhibit neurite outgrowth, we found that 1–1000 μg/ml CSPG added in solution failed to reach our criteria levels for significant elevation of $[Ca^{2+}]_i$ (Fig. 4C; mean $\Delta[Ca^{2+}]_i$ over all concentrations was 14.4 nM; the experiment was performed three times, each experiment testing the range 1–1000 μg/ml CSPG; $P > 0.05$). Importantly, these assays were done only on laminin. Whether the addition of soluble CSPG to DRGs elongating on fibronectin elevates $[Ca^{2+}]_i$ needs to be determined.

CSPG-induced elevation of $[Ca^{2+}]_i$ results in two types of recovery phases

Three types of responses occurred for DRG neurons that contacted CSPG, using either the CSPG bead or stripe techniques: (1) no elevation of $[Ca^{2+}]_i$ (17%); (2) elevation of $[Ca^{2+}]_i$ where recovery was fast (<5 min; mean = 3 min 43 s; $n = 8$), and rapidly returned to the original baseline (36.8%; $n = 19$) (Fig. 4D, closed circles); and (3) elevation of $[Ca^{2+}]_i$ where recovery was slow (>10 min; mean = 23 min 20 s; $n = 3$), and returned to a level that was higher than the original baseline (63.2%, $n = 19$) (Fig. 4D, open circles). Calcium transients often occurred during the slow recovery phase [109,110] (Fig. 4D, closed circles). These results may reflect the heterogeneity of DRGs, the past history of individual cells, or the amplitude of the initial calcium increase.

Implications for this heterogeneity of response may relate to the behaviour of DRG growth cones elongating on a CSPG step gradient [54], where sensory neurons decrease their rate of elongation with increasing concentrations of CSPG but are able to grow on to higher concentrations of CSPG than if they are not first adapted to lower levels, such as in the single stripe assay. For example, when a growth cone contacts the first and lowest concentration of CSPG in the step gradient, it may elevate $[Ca^{2+}]_i$ to a given level. Then, upon contact with a higher concentration of CSPG in the next tier of the step gradient, the degree to which $[Ca^{2+}]_i$ is elevated is less than it would be if the growth cone touched this concentration of CSPG directly from laminin. This allows the growth cone to continue to migrate, albeit more slowly. At some point specific to each neuronal type, the elevation of $[Ca^{2+}]_i$ may disallow further migration, facilitating growth cone turning.

CSPG elevates intracellular calcium in a guidance assay

Results presented above show that $[Ca^{2+}]_i$ was significantly elevated when CSPG-coated beads were touched to either cell bodies or growth cones, suggesting a possible signalling role for cytoplasmic calcium following CSPG contact. Since growth cones cease forward migration and/or turn in a guidance assay in which they encounter stripes of substratum-bound CSPG [47], we used this same paradigm to investigate whether contact with CSPG stripes results in changes in $[Ca^{2+}]_i$ in growth cones in this guidance assay.

For this study the 'CSPG stripe technique' was used (Fig. 1B). Bovine nasal cartilage CSPG (aggrecan-like), chick limb bud CSPG, or bovine aorta CSPG (versican-like), each as a starting solution of 1 mg/ml, was mixed with 5–10% rhodamine isothiocyanate (RITC) as a marker, and adsorbed to air-dried laminin-coated coverslips, three stripes per dish (transfer for approx. 5 min), using cellulose paper strips (350 μm × 10–20 mm), according to the method of Lagenauer and Lemmon [111]. For a complete description of the [CSPG] bound in this paradigm, see [47,54]. Following one wash with Hepes-buffered F14, dissociated DRGs were grown on the patterned substrata (see cell culture details above). Cytoplasmic calcium levels were measured as described above as DRG growth cones migrated first on laminin and then contacted the CSPG-adsorbed stripes.

Two approaches were used to examine changes in $[Ca^{2+}]_i$ when DRG growth cones contacted bound CSPG in the stripe assay. First, we measured changes in $[Ca^{2+}]_i$ for individual growth cone contacts as they occurred. Secondly, we analysed static images of many DRG neurons at various stages of contact with CSPG by measuring the $[Ca^{2+}]_i$ of growth cones at a CSPG stripe at low magnification.

In the first approach, growth cones of neurites extended from DRG neurons were observed in the CSPG stripe assay as they approached the border between laminin and CSPG at approximately right angles. When growth cones contacted CSPG, two changes were commonly observed: (1) some filopodia transiently retracted (partially or completely), but growth cones did not collapse; and (2) $[Ca^{2+}]_i$ was transiently elevated as in the CSPG bead assays described above. Elevation of $[Ca^{2+}]_i$ occurred in 83.3% ($n = 22$) of the growth cones that contacted CSPG-adsorbed stripes (mean $\Delta[Ca^{2+}]_i = 594 \pm 102$ nM; $P < 0.0001$; Fig. 3B). Growth cone $[Ca^{2+}]_i$ returned towards baseline levels following CSPG contact. In all cases, controls in which growth cones elongated from laminin on to laminin-adsorbed stripes without CSPG did not result in elevation of $[Ca^{2+}]_i$, nor did growth cones change their migratory behaviour [111a].

In the second approach, $[Ca^{2+}]_i$ was measured for growth cones either: (1) on laminin alone, or (2) at a border between laminin and CSPG at an undetermined time after initial contact with CSPG. On laminin alone, 18 out of 19 DRG neurons were at baseline $[Ca^{2+}]_i$ (160 ± 8.9 nM). In comparison, of the 41 neurons observed at the laminin/CSPG border, 50% were at baseline $[Ca^{2+}]_i$, 18% had a moderate elevation in $[Ca^{2+}]_i$ (<450 nM), and 32.0% had a high $[Ca^{2+}]_i$ (>700 nM). These results support our previous observations that a high $[Ca^{2+}]_i$ is not sustained following an initial increase, but $[Ca^{2+}]_i$ fluctuates and may return towards baseline over time. If this were not the case, we would have expected that more than 80% of the growth cones at a CSPG border would express high $[Ca^{2+}]_i$.

Growth cones routinely touch and pull away from a CSPG surface *in vitro* and most likely *in vivo* as well. In studies where calcium changes were monitored as they occurred, we have not determined whether growth cones that remain in continuous contact with CSPG repeatedly elevate $[Ca^{2+}]_i$. However, our observation of static images suggest that some form of adaptation occurs after initial contact with CSPG to then decrease $[Ca^{2+}]_i$ since 68% of the cell bodies and growth cones at a border were either at baseline or only moderate calcium elevation, and this percentage is too high to represent only those that are non-responders (approx. 12%).

CSPG contact does not result in growth cone collapse

Numerous studies describe a form of growth cone cessation of forward advance called growth cone collapse, i.e. a temporary or permanent reduction of the growth cone to a spindle shape with backward movement of the growth cone and axon [112–119]. However, growth cones that contact PGs do not show collapse above background levels [51]. Similarly, growth cone collapse does not occur when dissociated motor neurons contact cells from posterior sclerotome [120], a tissue rich in PGs [121]. Further, growth cone collapse is not always accompanied by elevated calcium [122]. In some instances, some filopodia do retract following contact with CSPG (Fig. 3B), but collapse most often does not occur (Fig. 3A). Furthermore, retraction of filopodia is a normal occurrence and is sometimes seen in response to control beads.

Although the mechanisms of inhibition involved when growth cone collapse occurs in response to other agents are unknown, differences between the results of other studies describing growth cone collapse in response to various non-PG inhibitors, and this study using CSPG suggest the following. (1) CSPG may represent a different mechanism of inhibition of growth cone migration to other agents, (2) the apparatus of growth cone motility is affected differently by CSPG than by other agents, and (3) calcium influx may be important, but as one of two or more components required to initiate collapse.

Source of $[Ca^{2+}]_i$ rise of CSPG-induced calcium increases in DRG neurons

The elevation of $[Ca^{2+}]_i$ that occurs when a growth cone encounters a stripe of CSPG is sufficient to cause major changes in growth cone behaviour. In order to determine whether elevation of $[Ca^{2+}]_i$ is necessary for the guidance provided by the CSPG, we first sought to identify the source(s) of Ca^{2+} that lead to the elevation of $[Ca^{2+}]_i$. Subsequently, we could perform experiments to block the rise in $[Ca^{2+}]_i$ and determine the resulting behaviour of DRG growth cones. Table 1 summarizes the results of these studies.

Potential sources of Ca^{2+} can be divided into two: those derived from influx of Ca^{2+} from the extracellular milieu, and those derived from intracellular calcium stores. Conditions and blockers, which have been extensively characterized in this and other systems and shown to alter the ability of cells to elicit changes in $[Ca^{2+}]_i$, were used in this study [109,123–125]. Our results show that influx of Ca^{2+} from extracellular sources appears to be required for the rise induced by CSPG. This influx is not likely to occur through conventional voltage-dependent channels. The ability of added nickel chloride, or lanthanum chloride, or the removal of extracellular calcium, to negate the CSPG-induced calcium rise does, however, demonstrate the importance of extracellular Ca^{2+} and, potentially, provide important tools for determining whether rises in calcium are necessary for providing guidance.

Table 1. **Treatments used to block elevation of $[Ca^{2+}]_i$ from either the extracellular environment or from intracellular stores**

Sources of Ca^{2+} reagent or method	Mean $\Delta[Ca^{2+}]_i$ (nM)	Controls (2 mM Ca^{2+}) (n)
Extracellular		
Voltage-gated channel blockers	+458 ± 153	3
General calcium-channel blockers	+498 ± 48	9
Nickel chloride	+16 ± 4	9
Lanthanum chloride	+29 ± 3	3
Zero calcium medium	+23 ± 10	5
Intracellular stores		
Dantrolene	+509 ± 55	5
Thapsigargin	+410 ± 58	3
Cyclopiazonic acid	+490 ± 71	3

Involvement of intracellular calcium stores was also examined using dantrolene, which blocks the ryanodine-sensitive calcium channel [126], and inhibitors of the microsomal Ca^{2+}-ATPase pump, thapsigargin (TG; irreversible) [126a], and cyclopiazonic acid (CPA; reversible) [127]. Addition of these reagents to DRGs in calcium-containing medium led to a large release of intracellular calcium that did not occur subsequently if cells were bathed in calcium-free medium to prevent refilling of the stores. These agents did not block CSPG-induced elevation of $[Ca^{2+}]_i$.

Growth cone avoidance of bound CSPG may not be dependent upon elevation of growth cone $[Ca^{2+}]_i$

Since we observed that medium lacking calcium, or medium to which nickel chloride or lanthanum chloride had been added, blocked the elevation of $[Ca^{2+}]_i$ following contact with CSPG, we examined, in particular, the effects of these treatments using the CSPG stripe behavioural assay for CSPG-induced guidance of DRG growth cones. In a series of experiments, DRG growth cone encounters with CSPG were analysed in the presence of either zero calcium medium, or media containing nickel chloride (10 µM–10 mM) or lanthanum chloride (1 mM), all of which are conditions or reagents that block elevation of $[Ca^{2+}]_i$ induced by CSPG. Furthermore, in order to compare our results with those of Bandtlow et al. [128], we also included in the study culture dishes treated with dantrolene (40 µM), TG (1 µM) and CPA (1 µM), which perturb intracellular stores. Each of the above reagents were added at the time of culturing DRG neurons. The cultures were analysed 12–14 h later using a × 20 phase-contrast objective. In all cases, growth cones did not elongate on to stripes adsorbed with CSPG, as determined by qualitative examination of DRG cultures in which CSPG stripes were labelled with the fluorescent marker RITC that clearly delineates the border between the CSPG-adsorbed stripe and laminin-coated regions.

Although these initial studies do not show that CSPG-induced elevation of $[Ca^{2+}]_i$ is necessary for the CSPG avoidance behaviour, these experiments are as yet preliminary. To assure that our treatments were effective in blocking the elevation of $[Ca^{2+}]_i$ throughout the course of the behaviour experiments, we must monitor $[Ca^{2+}]_i$ for prolonged periods of outgrowth, as well as document the detailed behaviour of growth cones under these conditions. Furthermore, we are undertaking a detailed analysis of growth cone behaviour such as filopodial lifetimes, rates of lamellipodial expansion and neurite outgrowth, and degree of adhesion to the substratum, during normal encounters with CSPG and when the elevation of $[Ca^{2+}]_i$ is blocked.

It may turn out that growth cones still avoid a CSPG-containing substratum in the absence of an elevation of $[Ca^{2+}]_i$. If so, this may indicate that an additional effect of the CSPG is important in the growth cone avoidance response. Integrins play an important role in neurite outgrowth and the guidance of growth cones, including those of sensory neurons [4,5,129–133]. Previous evidence indicates that CSPG can weaken cell adhesion to adhesive components of the ECM [8,16,134]. The negative charge of the polysaccharide chains of surface-bound CSPG may interfere with integrin binding to the substratum-bound adhesive ligands fibronectin and laminin. This would comprise another way to promote growth cone turning at a CSPG boundary by a mechanism that is independent of CSPG-induced elevation of $[Ca^{2+}]_i$.

In addition to these immediate effects of CSPG on integrin function, CSPG may also exert long-range effects on expression of integrin genes. Results show that α3 integrin mRNA is up-regulated when DRG growth cones migrate on low concentrations of CSPG bound to laminin in a CSPG step gradient [135], and that this up-regulation may facilitate adaptation for growth cone migration in a complex extracellular environment. Together, these results indicate that CSPG may affect integrin function and cell behaviour.

Summary

The results of the present study, taken together, demonstrate that DRG neurons are capable of responding to contact with CSPG by elevation of $[Ca^{2+}]_i$. Whether DRG neurons contact a CSPG-adsorbed surface at the cell body or at the growth cone, the results are identical. Large increases in $[Ca^{2+}]_i$ in the order of five to seven times resting levels are routinely observed. These increases require the carbohydrate portion of CSPG since enzymic removal of the carbohydrate abolishes the elevation of $[Ca^{2+}]_i$. Furthermore, the rise occurs irrespective of whether surface-bound CSPG is brought to the neuron or the neuron is allowed to grow out and encounter CSPG. It is of interest that in previous guidance experiments [47], the degree of inhibition of growth cone migration was related to the amount of CSPG bound to a substratum. In the present experiments, there is also a clear relationship between the amount of contact with CSPG and the calcium response. These results taken together, and coupled with the magnitude and time course of the calcium increases we have observed, show that PGs can induce an increase in $[Ca^{2+}]_i$ that could clearly provide a signalling function in DRG neurons, and may represent one mechanism by which PGs regulate growth cone turning at borders between growth-promoting and growth-inhibiting factors *in vivo*.

We would like to thank Dr. Jean Challacombe and Eric Brown for critical reading of the manuscript, and acknowledge the support of NIH grants EY01775, and EY10545 (to D.M.S.), HD19950 (to P.C.L.), and NS28323 (to S.B.K.), and an American Paralysis Association grant LB2-9402 (to D.M.S. and P.C.L.).

References

1. Doherty, P., Ashton, S.V., Moore, S.E. and Walsh, F.S. (1991) Cell **67**, 21–33
2. Bixby, J.L. and Harris, W.A. (1991) Annu. Rev. Cell Biol. **7**, 117–159
3. Bixby, J.L. (1989) Neuron **3**, 287–297
4. Tomaselli, K.J., Neugebauer, K.M., Bixby, J.L., Lilien, J. and Reichardt, L.F. (1988) Neuron **1**, 33–43
5. Tomaselli, K.J., Doherty, P., Emmett, C.J. et al. (1993) J. Neurosci. **13**, 4880–4888
6. Bixby, J.L., Grunwald, G.B. and Bookman, R.J. (1994) J. Cell Biol. **127**, 1461–1475
7. Luo, Y. and Raper, J.A. (1994) Curr. Opin. Neurobiol. **4**, 648–654
8. Rouslahti, E. (1988) Annu. Rev. Cell Biol. **4**, 229–255
9. Jackson, R.I., Busch, S.J. and Cardin, A.D. (1991) Physiol. Rev. **71**, 481–539
10. McKeon, R.J., Schreiber, R.C., Rudge, J.S. and Silver, J. (1991) J. Neurosci. **11**, 3398–3411
11. Canning, D.R., McKeon, R.J., DeWitt, D.A. et al. (1993) Exp. Neurol. **124**, 289–298
12. Levine, J.M. (1994) J. Neurosci. **14**, 4716–4730
13. Steindler, D.A., Settles, D., Erickson, H.P. et al. (1995) J. Neurosci. **15**, 1973–1983
14. Thomas, L.B. and Steindler, D.A. (1995) Neuroscientist **1**, 142–154
15. Funderburgh, J.L., Caterson, B. and Conrad, G.W. (1987) J. Biol. Chem. **262**, 11634–11640
16. Gallagher, J.T. (1989) Curr. Opin. Cell Biol. **1**, 1201–1218
17. Schwartz, N.B. and Smalheiser, N.R. (1989) in Neurobiology of Glycoconjugates (Margolis, R.U. and Margolis, R.K., eds.), pp. 151–186, Plenum Press, New York
18. Stallcup, W.B., Dahlin, K. and Healy, P. (1990) J. Cell Biol. **111**, 3177–3188
19. D'Angelo, M. and Greene, R.M. (1991) Dev. Biol. **145**, 374–378
20. Wight, T.N., Kinsella, M.G. and Qwarnstrom, E.E. (1992) Curr. Opin. Cell Biol. **4**, 793–801
21. Herndon, M.E. and Lander, A.D. (1990) Neuron **4**, 949–961
22. Noonan, D.M., Fulle, A., Valente, P. et al. (1991) J. Biol. Chem. **266**, 22939–22947
23. Rauch, U., Karthikeyan, L., Maurel, P., Margolis, R.U. and Margolis, R.K. (1992) J. Biol. Chem. **267**, 19536–19547
24. Carey, D.J., Evans, D.M., Stahl, R.C. et al. (1992) J. Cell Biol. **117**, 191–201
25. Kallunki, P. and Tryggvason, K. (1992) J. Cell Biol. **116**, 559–571
26. Tang, L.-H., Rosenberg, L., Reiner, A. and Poole, A.R. (1979) J. Biol. Chem. **254**, 10523–10531
27. Caterson, B., Christner, J.E., Baker, J.R. and Couchman, J.R. (1985) Fed. Proc. Fed. Am. Soc. Exp. Biol. **44**, 386–393
28. Heinegard, D. and Oldberg, A. (1989) FASEB J. **3**, 2044–2051
29. Margolis, R.K. and Margolis, R.U. (1989) Dev. Neurosci. **11**, 276–288

30. Davis, G.E., Klier, F.G., Engvall, E. et al. (1987) Neurochem. Res. **12**, 909–921
31. Hantaz-Ambroise, D., Vigny, M. and Koenig, J. (1987) J. Neurosci. **7**, 2293–2304
32. Walicke, P. (1988) Exp. Neurol. **102**, 144–148
33. Sheppard, A.M., Hamilton, S.K. and Pearlman, A.L. (1991) J. Neurosci. **11**, 3928–3942
34. LaFont, F., Rouget, M., Triller, A., Prochiantz, A. and Rousselet, A. (1992) Development **114**, 17–29
35. Wang, L. and Denburg, J.L. (1992) Neuron **8**, 701–714
36. Bovolenta, P., Fernaud-Espinosa, I. and Nieto-Sampedro, M. (1993) Soc. Neurosci. Abstr. **19**, 435
37. Halfter, W. (1993) J. Neurosci. **13**, 2863–2873
38. Bicknese, A.R., Sheppard, A.M., O'Leary, D.D.M. and Pearlman, A.L. (1994) J. Neurosci. **14**, 3500–3510
39. Faissner, A., Clement, A., Lochter, A. et al. (1994) J. Cell Biol. **126**, 783–799
40. McAdams, B.D. and McLoon, S.C. (1995) J. Comp. Neurol. **352**, 1–13
41. Ring, C., Lemmon, V. and Halfter, W. (1995) Dev. Biol. **168**, 11–27
42. Tuttle, R., Schlaggar, B.L., Braisted, J.E. and O'Leary, D.D.M. (1995) J. Neurosci. **15**, 3039–3052
43. Carbonetto, S., Gruver, M.M. and Turner, D.C. (1983) J. Neurosci. **3**, 2324–2335
44. Akeson, R. and Warren, S.L. (1986) Exp. Cell Res. **162**, 347–362
45. Verna, J.-M., Fichard, A. and Saxod, R. (1989) Int. J. Dev. Neurosci. **7**, 389–399
46. Snow, D.M., Steindler, D.A. and Silver, J. (1990) Dev. Biol. **138**, 359–376
47. Snow, D.M., Lemmon, V., Carrino, D.A., Caplan, A.I. and Silver, J. (1990) Exp. Neurol. **109**, 111–130
48. Cole, G.J. and McCabe, C.F. (1991) Neuron **7**, 1007–1018
49. Fichard, A., Verna, J.-M., Olivares, J. and Saxod, R. (1991) Dev. Biol. **148**, 1–9
50. Oohira, A., Matsui, F. and Katoh-Semba, R. (1991) J. Neurosci. **11**, 822–827
51. Snow, D., Watanabe, M., Letourneau, P.C. and Silver, J. (1991) Development **113**, 1473–1485
52. Brittis, P.A., Canning, D.R. and Silver, J. (1992) Science **255**, 733–736
53. Dou, C. and Levine, J.M. (1992) Soc. Neurosci. Abstr. **18**, 959
54. Snow, D.M. and Letourneau, P.C. (1992) J. Neurobiol. **23**, 322–336
55. Pindzola, R.R., Doller, C. and Silver, J. (1993) Dev. Biol. **156**, 34–48
56. Dou, C.-L. and Levine, J.M. (1994) J. Neurosci. **14**, 7616–7628
57. Friedlander, D.R., Milev, P., Karthikeyan, L. et al. (1994) J. Cell Biol. **125**, 669–680
58. Milev, P., Friedlander, D.R., Sakurai, T. et al. (1994) J. Cell Biol. **127**, 1703–1715
59. Oakley, R.A., Lasky, C.J., Erickson, C.A. and Tosney, K.W. (1994) Development **120**, 103–114
60. Snow, D.M., Atkinson, P., Hassinger, T., Kater, S.B. and Letourneau, P.C. (1994) Dev. Biol. **166**, 87–100
61. Perris, R. and Johansson, S. (1987) J. Cell Biol. **105**, 2511–2521
62. Perris, R. and Johansson, S. (1990) Dev. Biol. **137**, 1–12
63. Knox, P. and Wells, P. (1979) J. Biol. Chem. **40**, 77–88
64. Rich, A.M., Pearlstein, E., Weissmann, G. and Hoffstein, S.T. (1981) Nature (London) **293**, 224–226
65. Lewandowska, K., Choi, H.U., Rosenberg, L.C., Zardi, L. and Culp, L.A. (1987) J. Cell Biol. **105**, 1443–1454
66. Hassell, J.R., Robey, P.G., Barrach, H.-J. et al. (1980) Proc. Natl. Acad. Sci. U.S.A. **77**, 4494–4498
67. Zaremba, S., Guimaraes, S., Kalb, R.G. and Hockfield, S. (1989) Neuron **2**, 1207–1219
68. Nishiyama, A., Dahlin, K., Prince, J.T., Johnstone, S.R. and Stallcup, W.B. (1991) J. Cell Biol. **114**, 359–371
69. Rauch, U., Gao, P., Janetzko, A. et al. (1991) J. Biol. Chem. **266**, 14785–14801
70. DeWitt, D.A., Silver, J., Canning, D.R. and Perry, G. (1993) Exp. Neurol. **121**, 149–152
70a. Snow, D.M., Brown, E.M. and Letourneau, P.C. (1996) Int. J. Neurosci., in the press
71. Ernst, H., Zanin, M.K.B., Everman, D. and Hoffman, S. (1996) J. Cell Sci., in the press
72. Mattson, M.P. and Kater, S.B. (1987) J. Neurosci. **7**, 4034–4043
73. Kater, S.B. and Mills, L.R. (1991) J. Neurosci. **11**, 891–899
74. Cypher, C. and Letourneau, P.C. (1992) Curr. Opin. Cell Biol. **4**, 4–7
75. Goldberg, D.J. (1988) J. Neurosci. **8**, 2596–2605

76. McCaig, C.D. (1989) J. Cell Sci. **93**, 723–730
77. Rehder, V. and Kater, S. (1992) J. Neurosci. **12**, 3175–3186
78. Zheng, J.Q., Felder, M., Connor, J.A. and Poo, M. (1994) Nature (London) **368**, 140–144
79. Silver, R.A., Lamb, A.G. and Bolsover, S.R. (1989) J. Neurosci. **9**, 4007–4020
80. Lankford, K.L. and Letourneau, P.C. (1989) J. Cell Biol. **109**, 1229–1243
81. Luo, Y., Raible, D. and Raper, J. (1993) Cell **75**, 217–227
82. Doherty, P. and Walsh, F.S. (1994) Curr. Opin. Neurobiol. **4**, 49–55
83. Serafini, T., Kennedy, T.E., Galko, M.J. et al. (1994) Cell **78**, 409–424
84. Doege, K.J., Sasaki, M., Kimura, T. and Yamada, Y. (1991) J. Biol. Chem. **266**, 894–902
84a. Rosenberg, L.C., Tang, L.-H., Choi, L., et al. (1983) in Limb Development and Regeneration (Kelly, R.O., Goetnick, P.F. and MacCabe, J.A., eds.), pp. 67–84, A.R. Liss, New York
84b. Carrino, D.A. and Caplan, A.I. (1985) J. Biol. Chem. **260**, 122–127
85. Zimmerman, D.R. and Rouslahti, E. (1989) EMBO J. **8**, 2975–2981
86. Grynkiewicz, G., Poenie, M. and Tsien, R.Y. (1985) J. Biol. Chem. **260**, 3440–3450
87. Davenport, R.W. and Kater, S.B. (1992) Neuron **9**, 405–416
87a. Kater, S.B. and Mills, L.R. (1991) J. Neurosci. **11**, 891–899
88. Lankford, K.L. and Letourneau, P.C. (1991) Cell Motil. Cytoskel. **20**, 7–29
89. Letourneau, P.C. and Cypher, C. (1991) Cell Motil. Cytoskel. **20**, 267–271
90. Lankford, K.L., Cypher, C. and Letourneau, P.C. (1990) Curr. Opin. Cell Biol. **2**, 80–85
91. Letourneau, P.C. and Shattuck, T.A. (1989) Development **105**, 505–519
92. Janmey, P.A. (1994) Annu. Rev. Physiol. **56**, 169–191
93. Burridge, K. and Feramisco, J.R. (1982) Cold Spring Harbor Symp. Quant. Biol. **46**, 587–597
94. Blanchard, A., Ohanian, V. and Critchley, D.R. (1989) J. Muscle Res. Cell Motil. **10**, 280–289
95. Hartwig, J.H. and Kwiatowski, D.J. (1991) Curr. Opin. Cell Biol. **3**, 87–97
96. Weeds, A.G., Gooch, J., Hawkins, M., Pope, B. and Way, M. (1991) Biochem. Soc. Trans. **19**, 1016–1020
97. Yin, H.L. (1988) Bioassays **7**, 176–179
98. Challacombe, J.F., Snow, D.M. and Letourneau, P.C. (1995) Soc. Neurosci. Abstr. **13**
99. Hyman, C. and Pfenninger, K.H. (1987) J. Neurosci. **7**, 4076–4083
100. Siman, R. and Noszek, J.C. (1988) Neuron **1**, 279–287
101. Kennedy, M.B. (1989) Trends Neurosci. **12**, 417–420
102. Hinrichsen, R.D. (1993) Biochim. Biophys. Acta **1155**, 277–293
103. VanBerkum, M.F.A. and Goodman, C.S. (1995) Neuron **14**, 43–56
104. Gordon, H., Lupa, M., Bowen, D. and Hall, Z. (1993) J. Neurosci. **13**, 586–595
105. Hockfield, S., Kalb, R.G., Zaremba, S. and Fryer, H. (1990) Cold Spring Harbor Symp. Quant. Biol. **1990**, 505–514
106. Kuhn, T.B., Schmidt, M.F. and Kater, S.B. (1995) Neuron **14**, 275–285
107. Reference deleted.
108. Snow, D.M. (1994) Am. Soc. Cell Biol. **34**, 364a
109. Tsien, R.W. and Tsien, R.Y. (1990) Annu. Rev. Cell Biol. **6**, 715–760
110. Gomez, T.M., Snow, D.M. and Letourneau, P.C. (1995) Neuron **14**, 1233–1246
111. Lagenaur, C. and Lemmon, V. (1987) Proc. Natl. Acad. Sci. U.S.A. **84**, 7753–7757
111a. Gomez, T.M. and Letourneau, P.C. (1994) J. Neurosci. **14(10)**, 5959–5972
112. Kapfhammer, J. and Raper, J. (1987) J. Neurosci. **7**, 201–212
113. Walter, J., Henke-Fahle, S. and Bonhoeffer, F. (1987) Development **101**, 909–913
114. Schwab, M.E. and Caroni, P. (1988) J. Neurosci. **8**, 2381–2393
115. Cox, E.C., Mueller, B. and Bonhoeffer, F. (1990) Neuron **4**, 31–37
116. Davies, J.A., Cook, G.M.W., Stern, C.D. and Keynes, R.J. (1990) Neuron **4**, 11–20
117. Muller, B., Stahl, B. and Bonhoeffer, F. (1990) J. Exp. Biol. **153**, 29–46
118. Raper, J.A. and Kapfhammer, J.P. (1990) Neuron **4**, 21–29
119. Johnston, A.R. and Gooday, D.J. (1991) Development **113**, 409–417
120. Oakley, R.A. and Tosney, K.W. (1993) J. Neurosci. **13**, 3773–3792
121. Tosney, K.W. and Landmesser, L.T. (1985) J. Neurosci. **5**, 2345–2358
122. Ivins, J.K., Raper, J.A. and Pittman, R.N. (1991) J. Neurosci. **11**, 1597–1608
123. Fox, A.P., Nowycky, M.C. and Tsien, R.W. (1987) J. Physiol. **394**, 149–172
124. Thastrup, O., Cullen, P.J., Drobak, B.K. and Hanley, M.R. (1990) Proc. Natl. Acad. Sci. U.S.A. **87**, 2466–2470

125. Foskett, J.K. and Wong, D. (1992) Am. J. Physiol. **262**, C656–C663
126. Palade, P., Dettbarn, C., Alderson, B. and Volpe, P. (1989) Mol. Pharmacol. **36**, 673–680
126a. Thastrup, O., Cullen, P.J., Drobak, B.K. and Hanley, M.R. (1990) Proc. Natl. Acad. Sci. U.S.A. **87**, 2466–2470
127. Holzapfel, C.W. (1968) Tetrahedron **24**, 2101–2119
128. Bandtlow, C.E., Schmidt, M.F., Hassinger, T.D., Schwab, M.E. and Kater, S.B. (1993) Science **259**, 80–83
129. Letourneau, P.C., Pech, I.V., Rogers, S.L. et al. (1988) J. Neurosci. Res. **21**, 286–297
130. Neugebauer, K.M., Tomaselli, K.J., Lilien, J. and Reichardt, L.F. (1988) J Cell Biol. **107**, 1177–1187
131. Lein, P.J., Higgins, D., Turner, D.C., Flier, L.A. and Terranova, V.P. (1991) J. Cell Biol. **113**, 417–428
132. Haugen, P.K. (1992) J. Neurosci. **12**, 2034–2042
133. Venstrom, K. and Reichardt, L. (1995) Mol. Biol. Cell. **6**, 419–431
134. Culp, L.A., Laterra, J., Lark, M.W., Beyth, R.J. and Tobey, S.L. (1986) in Functions of the Proteoglycans, Ciba Geigy Foundation, vol. 124, pp. 158–183, Wiley, Chichester
135. Condic, M.L., Snow, D.M. and Letourneau, P.C. (1993) Soc. Neurosci. Abstr. **19**, 436

7

Mechanisms of growth cone collapse by central nervous system myelin-associated neurite growth inhibitors

Christine E. Bandtlow

Brain Research Institute, University of Zurich, August-Forel-Str. 1,
8029 Zurich, Switzerland

Introduction

The functioning of the nervous system of higher vertebrates depends on the underlying detailed and highly stereotyped patterns of neuronal connectivity. The specificity of synaptic connections is accomplished during development in three major steps: pathway selection, recognition of the target region and formation and stabilization of specific connections (address selection). The distal tip of the extending neurite, termed the growth cone [1], is a highly motile structure and plays an important role in determining the direction and distance of neuronal growth to find the correct target. As they advance, growth cones are exposed to a series of extrinsic guidance cues to which they can respond. Thus they navigate along specific pathways in a remarkably unerring way. Once they reach their proper targets neurite elongation stops and hypothetical recognition molecules lead to the formation of synapses. In this way, a first scaffold of projections and synapses is established in the early nervous system which allows later outgrowing fibres to follow. The initial steps of growth cone guidance typically occur before neurons become functionally active and rely on molecular mechanisms of pathway and target recognition which are largely activity independent. These initial patterns of connections are then refined, as axonal terminals retract and re-expand to select specific subsets of cells in the target region. This remodelling, called address selection, is the consequence of competition with surrounding inputs and allows the transformation of a coarse and imprecise projection into a refined and highly tuned pattern of functional connections. In recent years it has become more and more apparent that this process of activity-dependent synaptic remodelling does not stop at birth but continues throughout the lifetime of an organism and accounts for plasticity, learning and memory.

Since the original proposal of Sperry [2] that the 'homing behaviour' of growth cones can be explained by 'differential chemical attraction' many attempts have been made to uncover the nature and function of the proposed recognition molecules. A number of *in vitro* studies have suggested that differential adhesiveness may play a role in growth cone guidance [3] and that differential expression of cell adhesion molecules (CAMs) or extracellular matrix (ECM) material may be responsible for axonal guidance.

A significant number of CAMs and ECM constituents have been characterized on the molecular level, many of which are expressed in the developing nervous system [4] and the number of novel molecules is still increasing. A large number of these neural CAMs in both vertebrates and invertebrates are members of two different gene families, the immunoglobulin superfamily [4,5] and the cadherin family [6], although other families exist as well [7,8]. Several mechanisms have been identified by which CAMs interact with other cell-surface components: homophilic binding between identical molecules on opposing cell surfaces, heterophilic binding and assisted homophilic binding between different CAMs or complexes of two CAMs [5]. Moreover, a diverse family of receptors of the integrin type have been discovered on growth cones as the binding molecules for ECM molecules [9,10]. A common property of many CAMs and integrin-type receptors is not only their restricted expression on subsets of axons but also their regulated spatial-temporal expression during development. Thus axons can change their repertoire of CAMs or integrins when switching from one microenvironment to another. This implies that the neuronal requirements for and responsiveness to different CAMs and ECM constituents (and guidance factors in general) changes during ontogeny.

To what extent individual CAMs and ECM molecules play a crucial or decisive role in axonal pathfinding *in vivo* is still a largely open question. It is clear that many CAMs, integrins and ECM molecules have well-characterized effects (mainly neurite growth-promoting) on growing neurites in culture, and that they show specific expression patterns *in vivo*. However, in only a few cases has their function in the developing organism been successfully demonstrated [11]. While such molecules are likely to provide substrates for migrating neurons and extending axons, gene deletion experiments have demonstrated that the absence of a single molecule is often reflected in only minor consequences for cell migration, axon extension and the formation of a functional nervous system [12–14]. Such astonishing results are interpreted as a reflection of 'redundancy' in the molecular orchestra responsible for axonal guidance and indicate that only the combinatorial interaction of several CAMs, ECM constituents and integrins brings about guidance of particular fibre tracts. First confirmations of this hypothesis have already been provided by the loss-of-function studies of multiple CAMs [15,16].

Besides membrane-bound molecules supporting or stimulating neurite growth, a number of soluble, diffusible factors affecting growth cone behaviour have been found. That nerve fibres could be guided by gradients of diffusible signals emanating from their target cells was originally proposed by Ramon y Cajal [1]. In the meanwhile, quite a few factors isolated from various target tissues have been described with chemoattractive and repulsive activities for appropriate growth cones [17–23]. Moreover, previous studies on isolated cultured neurons suggested that growth cone behaviour can be influenced by electrical activity and certain neurotransmitters [24–26], implying that the local release of neurotransmitters or the electric activity of target neurons may play a role in synapse formation and sprouting.

Besides extension-promoting cues, a growth cone is also exposed to negative guidance cues. Much less information is available on this class of recently discovered guidance factors, but the interest in the presence of repulsive or inhibitory signals is increasing [23,27]. One of the first hints for the importance of negative guidance cues was the finding that peripheral axons in chick embryos are growing through a narrow channel of non growth-supportive tissue, the developing chondrocytes that form the pelvic girdle [28,29]. Similarly, during early chick development, motor neurons extend their axons only through the anterior part of the somites but actively avoid the posterior half. Keynes, Stern and colleagues have shown that motor axons still prefer the anterior to the posterior portion when somites are surgically displaced [30,31]. Using a growth cone collapse assay it was shown that membranes from posterior somites, but not those from anterior somites, induce growth cone collapse [32], suggesting that the *in vivo* patterning may arise by repulsion of motor axons from posterior half-somites. Interestingly, this activity co-purifies with biochemical fractions containing two conspicuous posterior somite-specific components of 48 kDa and 55 kDa. Unfortunately, complete purification and molecular cloning of these molecules has not yet been achieved.

A well-characterized system where nerve fibres show specific topographic preferences for their target areas is the retino-tectal system. During chick development retinal axons grow to their target, the optic tectum, where they form a topographic map, i.e. temporal retinal axons terminate in the anterior half of the tectum, nasal axons in the posterior half [33]. Bonhoeffer and colleagues have shown in a number of *in vitro* assays that this preference is due in part to a nerve-fibre-repulsive activity present in the membranes of posterior tectum [34,35]. By examining the choices that temporal fibres make between membranes isolated from successive rostro-caudal parts of the tectum it became clear that this inhibitory activity is graded over the caudal two-thirds of the tectum [35]. Temporal axons challenged *in vitro* with a steep gradient of the posterior activity responded in a graded manner to the increasing concentrations [36]. Recent purification and biochemical characterization of this inhibitory activity derived from posterior tectal membranes indicated that it is associated with a glycolipid-anchored protein of an apparent molecular mass of approx. 25 kDa and consistent with the hypothesis of a repulsive gradient; *in situ* hybridization for the presence of this protein in the chick tectum reflects a gradient with increasing intensity towards the posterior tectal pole [36a].

In vitro studies have shown several examples of interactions that can be interpreted as repulsion or growth inhibition. Thus, central nervous system (CNS) growth cones avoid peripheral nervous system (PNS) axons, and vice versa [37–39], nasal retinal axons avoid temporal retinal axons [40], and motor axons avoid sensory axons [41]. Not only do growth cones of one cell type avoid the axons of another neuron class, but upon contact the growth cones collapse and retract. Based on the observation that PNS growth cones are inhibited upon contact with CNS fibres, Raper and colleagues have purified a collapse-inducing activity from chick brain. Recently, the successful identification and cloning of a glycoprotein of 110 kDa has been reported which induces growth cone collapse and was termed collapsin [42]. Interestingly, the N-terminal half of collapsin shares significant sequence identity with fasciclin IV, an immunoglobulin superfamily membrane protein that has a role in growth cone guidance in grasshopper and *Drosophila* [43,44].

Evidence for the presence of neurite growth inhibitors in the adult CNS of higher vertebrates arose in the context of studies on the absence of fibre regeneration of lesioned CNS fibre tracts. Tissue-culture experiments revealed that the failure of nerve fibres to regrow through CNS tissue is not due to an intrinsic constraint of CNS neurons but at least in part to the presence of inhibitory molecules in the CNS [45]. It was initially shown that these neurite growth-inhibitory molecules are present on differentiated oligodendrocytes and in CNS myelin but are absent from PNS myelin [46,47]. Preliminary biochemical analysis revealed that the inhibitory activity of CNS myelin resides in two protein fractions migrating at approx. 35 kDa and 250 kDa on denaturing polyacrylamide gels [47]. According to their molecular masses these protein fractions were called NI-35 and NI-250. A monoclonal antibody (IN-1) directed against the 250 kDa protein fraction was able to neutralize the inhibitory activity associated with CNS myelin and of purified NI-35 and NI-250 [48]. That these myelin-associated neurite growth inhibitors play an important role for the absence of CNS fibre regeneration *in vivo* in the lesioned spinal cord, brain and optic nerve was demonstrated in a number of experiments where long-distance regeneration of lesioned fibre tracts was obtained in animals treated with the neutralizing IN-1 antibody [49–53]. In addition, fibre tract regeneration was also shown in spinal cords and optic nerves where myelination was experimentally prevented [52,54]. Recent results indicate that these myelin-associated inhibitors may also play a role in regulating sprouting and plasticity in the developing and normal adult CNS of higher vertebrates [55–58].

It is, therefore, reasonable to assume that repulsive and inhibitory molecules play equally important roles in axonal guidance as neurite growth-promoting factors. The development of the nervous system and its complex functional connectivity probably result from the temporally and spatially co-ordinated expression of positive as well as negative guidance signals.

While the *in vitro* effects of growth-promoting and repulsive/inhibitory molecules are often well described, the biochemical and molecular bases for these effects have remained largely elusive. Since fundamental decisions on neuronal morphology, axonal pathfinding and connectivity are made at the level of the neuronal growth cone, it is of great interest to learn not only about the many external signals impinging on the growth cone but also how these signals are received and integrated by the growth cone to determine its behaviour. In addition, the problem of how independence of several stimulus–response pathways is maintained when different stimuli share elements of the same signal transduction pathway is central to understanding of information processing in the nervous system.

Oligodendrocytes arrest neurite growth by contact inhibition

From the previous observations by Schwab and co-workers it became obvious that differentiated oligodendrocytes and their *in vivo* product myelin are inhibitory for growing neurons in culture [45,47,48]. In time-lapse video-microscopy individual growth cones of perinatal dorsal root ganglion (DRG) cells extending on laminin in the presence of nerve growth factor (NGF) (i.e. under optimal growth-promoting conditions) were observed as they encountered dissociated CNS glial cells from newborn rat optic nerve [59]. Growth cones of extending neurites considered astrocytes as favourable substrates with little or no effect on growth cone morphology and growth velocity. In contrast, growth cone contact with differen-tiated, galactocerebroside-positive oligodendrocytes resulted in a rapid and long-lasting arrest of growth cone movement, followed by collapse and retraction of the growth cone structure. In this respect it was interesting to observe that the contact of a few growth cone filopodia was sufficient to cause growth arrest [59]. Collapsed and retracted growth cones were able to recover to normal motility and they maintained their responsiveness to collapse upon contact with an oligodendrocyte. In order to investigate the possible involvement of the previously described growth inhibitory molecules NI-35 and NI-250 of CNS myelin [47,48] in this contact-mediated growth cone response, encounters between oligodendrocytes and growth cones were monitored in the presence of the neutralizing monoclonal antibody IN-1. Interestingly, in the presence of IN-1 antibodies growth cones were not arrested upon contact with oligodendrocytes and were able to grow over oligodendrocyte processes without morphological changes to their lamellipodial structure [59]. Several conclusions could be drawn from these experiments as follows.

(1) Contact inhibition was only seen with oligodendrocytes but not with astrocytes.

(2) Arrest and collapse of growth cones was likely to be evoked by the expression of the IN-1 antigen(s), since growth cones did not collapse but could grow over oligodendrocyte processes in the presence of the IN-1 antibody.

(3) Arrest and collapse of growth cones were strictly contact-mediated phenomena. A diffusible inhibitory factor secreted from oligodendrocytes could therefore be excluded.

(4) Growth inhibition was strictly local; growth cones of other branches of a collapsed neurite that were not in contact with an oligodendrocyte did not collapse.

(5) Interestingly, contact of a few growth cone filopodia was sufficient to induce collapse and retraction, indicating that filopodia express the appropriate receptors to detect the inhibitory molecules and that a second messenger can be efficiently transduced to the growth cone machinery.

Second messengers involved in growth cone collapse

Many of the underlying processes that may play roles in neurite outgrowth and growth cone motility can be significantly affected by varying the levels of intracellular calcium. Calcium-channel blockers or removal of extracellular calcium have been reported to inhibit neurite extension and alter growth cone morphology in several neuronal cell types [60,61]. In addition, calcium currents have been identified in actively growing neurites, but not in spontaneously inactive growth cones [62]. Previous results demonstrated that a variety of stimuli that can affect the behaviour of growth cones *in vitro* do so by changing the intracellular calcium. Mechanical stimuli, action potentials [24,63,64] and neurotransmitters [25,65–68] induce growth cone turning or collapse by increasing the intracellular calcium levels by influx of calcium through voltage-gated calcium channels. This has lead to the 'calcium optimum' hypothesis of growth cone motility, which proposes that neurite outgrowth is possible when the intracellular calcium concentration ($[Ca^{2+}]_i$) is maintained within a limited range [69,70]. Should the concentration of free calcium deviate above or below this permissive range, spontaneously or in response to specific stimuli, growth cone motility is arrested. A modification of this idea, the 'set-point' hypothesis, predicts that regulatory mechanisms can return the calcium concentration to this permissive range during maintained stimulation, to permit further outgrowth [70]. Although it was clear that certain stimuli affecting growth cone behaviour may well act through regulators other than intracellular calcium [71–73], the possible role for intracellular calcium in the regulation of growth cone behaviour claimed by the calcium hypothesis had to be examined in the context of growth cone collapse evoked by the myelin-associated neurite growth inhibitors NI-35/NI-250.

We made use of a bioassay which allows us to monitor for changes of intracellular calcium within a given growth cone and morphological or behavioural changes at the same time. For this purpose, fluorescence imaging using the calcium indicator fura-2/acetoxymethyl (AM) ester was done in combination with time-lapse phase-contrast microscopy to analyse whether changes in growth cone morphology were correlated with changes in $[Ca^{2+}]_i$. DRG growth cones were exposed to NI-35-containing liposomes locally applied with a micropipette. We could show that growth cones responded to NI-35 within seconds with a rapid collapse and retraction [74]. This growth cone response was very similar to the morphological changes seen in growth cones upon contact with oligodendrocytes [59]. The direct measurements of $[Ca^{2+}]_i$ revealed that the observed growth cone collapse was preceded by a large, transient rise in calcium [74]. Preincubation of NI-35-containing liposomes with IN-1 antibody (the neutralizing antibody) prevented this increase in calcium, as well as the subsequent morphological changes. To demonstrate that the increase in $[Ca^{2+}]_i$ was causally involved in inducing growth cone collapse, a variety of pharmacological reagents that block entry of extracellular calcium or perturb calcium release from intracellular Ca^{2+} stores were tested. Interestingly, voltage-gated calcium-channel blockers could not prevent the calcium rise or the morphological changes. However, dantrolene, an inhibitor of calcium release from caffeine/rhyanodine-sensitive intracellular calcium stores, protected growth cones from collapse in the presence of NI-35. Depletion of these calcium stores by repetitive pulses of caffeine drastically reduced the calcium increase normally induced by NI-35. These studies provided evidence in support of the hypothesis that an increase of the $[Ca^{2+}]_i$ within DRG growth cones is a crucial step in mediating the collapse response evoked by NI-35 [74].

Retinal axon growth cone responses to different inhibitory molecules can be mediated by calcium-dependent and -independent pathways

Evidence was provided above that the CNS-myelin-associated neurite growth inhibitor NI-35 mediates its growth-cone-collapsing effect in DRG neurons by the release of calcium from intracellular stores. We were interested to study the effect of NI-35 on other types of neurons and to investigate whether a given molecule (NI-35) induces the same morphological responses, using the same second messenger pathway. These questions were particularly interesting in the light of recent reports which have shown that a given stimulus can activate different responses in different cells. For example, when the neurotransmitter 5-hydroxytryptamine (serotonin) is applied to the *Helisoma* neuron B19, growth cones abruptly go through a series of morphological changes leading to growth cone collapse and inhibition of neurite elongation [75]. This effect is highly neuron-selective, since growth cones of neuron B5 are completely unaffected by the same concentration of 5-hydroxytryptamine. In addition, different stimuli evoking similar responses in a given cell may use different second messenger systems. Recently it has become apparent that various soluble or membrane-bound molecules can induce neuronal growth cone collapse by using different signal transduction pathways [71,72].

Two questions were addressed. First, do NI-35-containing liposomes induce collapse of growth cones of chick retina explants; and if so, is this response also mediated by calcium as seen for rat DRG cells? Secondly, does another growth-cone-collapsing activity, derived from chick posterior tectal membranes and known to evoke retinal growth cone collapse, use intracellular calcium as a second messenger?

In chick embryos, the post-mitotic ganglion cells in the retina send out their first axons at embryonic day (E) 3 towards the optic fissure where they enter the optic nerve, cross the optic chiasm and grow towards the tectum. Interestingly, a topographic projection is typical for this system: axons of temporal, nasal, dorsal and ventral retinal ganglion cells project to the anterior, posterior, ventral and dorsal tectum, respectively [76]. Different models have been discussed to explain this topographic projection of retino-tectal fibres [2]. Sperry's chemoaffinity model (1963) proposed gradients of matching substrate-bound attractants on both retinal axons and their tectal targets, whereas another model has been based on differential adhesion. However, the number of positional markers including their corresponding receptors required to explain the chemoaffinity model would probably go beyond the capacity of the genome [77].

To learn more about position-based guidance cues in the retino-tectal projection, Bonhoeffer and colleagues established a series of *in vitro* assays using chick retinal explants [34,35,78]. Ganglion cell axons from either temporal or nasal retina explants were given the choice to grow on alternating stripes of membranes prepared from either anterior or posterior tectum. The results of these experiments were quite striking. Whereas axons from nasal retina were able to grow on membranes from both anterior and posterior tectum, the temporal fibres showed a preference for growing on membrane stripes prepared from anterior tectum [34,35,78]. Further tests showed that the response of the temporal axons is based on avoidance of the posterior tectum, rather than simply attraction to the anterior tectum, since prior treatment of the posterior membranes with heat, protease, or phospholipase C changes them into a permissive substrate for temporal axons [34,35,79]. In addition, enrichment of certain membrane components of the anterior tectum resulted in a preferential outgrowth response of nasal axons on membrane stripes of their proper target [80]. The dynamic behaviour of retinal ganglion cell growth cones was monitored upon addition of posterior tectal membranes: posterior membrane vesicles induce a fast, dose-dependent collapse of temporal growth cones [81], whereas nasal growth cones collapsed but recovered within minutes. Biochemical and molecular characterization of the collapsing activity derived from chick posterior tectal membranes indicated that it was associated with a glycolipid-anchored protein of an apparent molecular mass of approx. 25 kDa, which is only expressed in the posterior part of the chick tectum [36a].

In collaboration with Bonhoeffer's group we have now investigated whether the myelin-associated and the posterior tectal membrane-derived activities use the same signal transduction pathway to evoke growth cone collapse. For reasons of material limitations, however, we used crude membrane fractions of chick E10 posterior tectum instead of the purified protein.

Retinal explants from E6/E7 chick embryos were plated on laminin-coated dishes and cultured in F12 medium overnight. After 12–18 h in culture extensive fibre outgrowth of nasal and temporal fibres was seen. As described above, individual growth cones were monitored under the microscope before and during addition of the collapse-inducing proteins.

Liposomes containing NI-35 (derived from rat spinal cord myelin) induced collapse and retraction in 85% of the observed growth cones of temporal or nasal growth cones. In contrast to rat DRGs, however, this collapse response was transient, since 75% of the collapsed growth cones recovered despite the presence of NI-35 and showed their normal motility within 10 min. This recovery effect was dose-independent. As seen for rat DRGs, collapse was not seen with control liposomes or with NI-35-containing liposomes which had been preincubated with the IN-1 antibody (see Table 1). Membranes of E10 chick posterior tectum (p-membranes) induced growth cone collapse in 100% of the observed temporal fibres, whereas nasal fibres showed an initial collapse response, but recovered within 1–2 min. This recovery response was clearly faster than that seen after treatment with NI-35-containing liposomes. Interestingly, whereas NI-35-evoked growth cone collapse was transient, collapse of temporal growth cones induced by p-membranes persisted for more than 6 h (Table 1).

Calcium measurements before and during application of NI-35-containing liposomes showed that growth cone collapse was preceded by a transient rise in intracellular calcium levels (Table 2). Interestingly, the presence of 20 μM dantrolene could prevent the morphological changes (Table 1), indicating a causal relationship between the calcium change and the collapse response. This response is very similar to the one seen in rat DRG growth cones, indicating that release of calcium from intracellular calcium stores seems to be a general mechanism mediating the morphological growth cone responses evoked by NI-35. Interestingly, however, whereas in DRG growth cones calcium levels were elevated within 30 s of NI-35-containing liposome addition, the peak levels for calcium in temporal or nasal retinal ganglion growth cones was only seen after approx. 4 min (Table 2). This effect could be explained by different NI-35-receptor concentrations on the various neuronal subtypes, the lack of an additional intracellular second messenger pathway, smaller intracellular calcium pools, potent and fast calcium pumps or other mechanisms operating in retinal ganglion cells, which could also explain the only transient collapse response of retinal ganglion cells to NI-35.

Although p-membranes induced a long-lasting collapse of temporal axons, no detectable changes of $[Ca^{2+}]_i$ could be measured (Table 2). In addition,

Table 1. Growth cone response of RGCs after treatment with various protein fractions

	Collapse (%)	Recovery (%)	Time until collapse (min)	Collapse duration (min)
P-membranes*	100 (n=47)	0 (n=47)	2.2 ± 0.2 (n=47)	> 6 h (n=47)
NI-35 liposomes	85 (n=54)	75 (n=52)	6.3 ± 0.5 (n=52)	10.4 ± 1.6 (n=52)
Control liposomes	17 (n=27)	–	–	–
NI-35 liposomes + IN-1	33 (n=21)	–	–	–
NI-35 liposomes + dantrolene	10 (n=88)	–	–	–
P-membranes + dantrolene	65 (n=20)	–	–	–

*Numbers for p-membranes are only for temporal retinal fibres.

Table 2. Intracellular calcium concentrations (in nM) before and after addition of various protein fractions

	Ca peak pre-addition (nM)	Ca peak post-addition (nM)	Time until peak (min)	Peak width (min)
P-membranes	109.6 ± 8.65 (n=56)	116.5 ± 8.55 (n=56)	–	–
NI-35 liposomes	126.3 ± 5.03 (n=32)	554.17 ± 57.43 (n ± 27)	3.89 ± 0.77 (n=27)	1.5 ± 0.16 (n=27)
Control liposomes	124.5 ± 7.07 (n=27)	179.0 ± 9.2 (n=27)	–	–
NI-35 liposomes + IN-1	121.7 ± 6.05 (n=29)	189.7 ± 9.8 (n=29)	–	–

65% of the temporal growth cones which no longer responded to NI-35 in the presence of 20 μM dantrolene, collapsed upon addition of p-membranes (Table 1).

Conclusion

These results suggest that NI-35 induces calcium release from caffeine/ryanodine-sensitive intracellular stores as a general mechanism for growth cone collapse, whereas p-membrane-evoked growth cone collapse of temporal retinal fibres does not seem to be correlated with any detectable changes of $[Ca^{2+}]_i$. At the moment, the second messenger(s) involved in this calcium-independent pathway used by posterior-tectal membrane factors remains unknown. The comparison of retinal ganglion cells and DRGs suggests that NI-35 induces an intracellular calcium rise by similar mechanisms in both neuronal cell types, namely by the release of calcium from intracellular stores. Given the accepted coupling of calcium release and influx [82,83] an additional role for calcium influx from the extracellular medium can not be ruled out.

Igarashi and co-workers [84] have shown that NI-35-induced growth cone collapse of embryonic chick DRGs is greatly reduced in the presence of pertussis toxin, suggesting the involvement of a G-protein-coupled receptor. The activation of such a receptor could result in the production of inositol-1,4,5-trisphosphate (InsP_3), a second messenger known to induce release of calcium from InsP_3-sensitive intracellular stores. Our own results, however, suggest that calcium is released from caffeine/ryanodine-sensitive and not from InsP_3-sensitive stores, since dantrolene specifically blocks the caffeine/ryanodine receptor. Until recently, the endogenous ligand for this calcium channel was not known. Increasing evidence now supports the idea that cyclic ADP-ribose, a metabolite of NAD, can act as an endogenous ligand of the caffeine/ryanodine receptor, thereby mobilizing the intracellular release of calcium [85,86]. In addition, an NAD-glycohydrolase was found to synthesize and degrade cyclic ADP-ribose [87]. NAD-glycohydrolases or cyclic ADP-ribose cyclases are ubiquitous membrane-bound enzymes that have been known for many years but whose function had not been identified. Sequence information of the *Aplysia* cyclic ADP-ribose cyclase revealed structural properties shared with CD-38, a phenotypic marker of different subpopulations of B- and T-lymphocytes. Murine CD38, a typical type-II membrane glycoprotein, that consists of a short cytoplasmic domain, a membrane-spanning domain and a large extracellular domain, was shown to exhibit ADP-ribosyl cyclase activity. Interestingly, the formation of cyclic ADP-ribose is enhanced by elevated cyclic GMP levels [85]. Cyclic ADP-ribose may therefore function as a calcium-mobilizing intracellular messenger for ligands activating guanylate cyclases.

On the basis of these results it is tempting to speculate that NI-35 is mediating its effect by the formation of cyclic ADP-ribose by the direct or indirect activation of a membrane-bound NAD-glycohydrolase. The production of cyclic

ADP-ribose would then induce the release of calcium from intracellular stores. Possible mechanisms by which calcium might alter growth cone morphology are by acting directly, or through kinases or phosphatases, on the regulation of cytoskeletal elements that underlie growth cone motility [88]. Several proteins known to affect actin stability have been demonstrated immunocytochemically in neuronal growth cones, including cross-linking proteins such as α-actinin and filamin, as well as membrane-anchoring proteins such as talin and vinculin. Moreover, actin-depolymerizing factor has been demonstrated to be a major component in the growth cone. Profilin, which is thought to regulate the levels of assembled actin, as well as villin and gelsolin are likely candidates to be involved in the regulation of the actin-based cytoskeleton [89–92]. The roles, however, that these proteins might play and the way they might interact within the growth cone remain to be understood.

Although there is ample evidence that elevation of $[Ca^{2+}]_i$ can cause growth cone collapse, not all calcium increases are associated with changes in growth cone morphology [93]. Electrical activity can increase growth cone calcium over time to levels greater than 500 nM but growth cones fail to collapse [63,94]. This may be due to a slow adaptation response of the growth cone to elevated calcium levels, when the calcium increase occurs with a slow time kinetic. This implies that not only the intensity or type of second messenger activated by a certain stimulus is of significance but that the kinetic properties of elements comprising signal transduction pathways confer sensitivity to specific patterns of activation.

I would like to thank M.E. Schwab for his continuous support and fruitful discussions, J. Löschinger for sharing his data and B. Niederöst for her powerful input in the protein purification of NI-35/250. This work was supported by the Swiss Natl. Sci. Found., Grant no. 31-42299.94.

References

1. Ramon y Cajal, S. (1959) Degeneration and Regeneration of the Nervous System, Hafner, New York
2. Sperry, R. (1963) Proc. Natl. Acad. Sci. U.S.A. **50**, 703–710
3. Liang, P. and Pardee, A.B. (1992) Science **257**, 967–971
4. Jessell, T.M. (1988) Neuron **1**, 3–13
5. Sonderegger, P. and Rathjen, G.F. (1992) J. Cell Biol. **119**, 1387–1394
6. Hynes, R.O. (1992) Curr. Opin. Gen. Dev. **2**, 621–624
7. Smalheiser, N. and Collins, R.B. (1992) J. Dev. Brain Res. **69**, 215–223
8. Smalheiser, N. and Collins, R.B. (1992) J. Dev. Brain Res. **69**, 225–231
9. Tomaselli, K.J., Doherty, P., Emmett, C. et al. (1993) J. Neurosci. **13**, 4880–4888
10. Reichardt, L.F. and Tomaselli, K.J. (1991) Annu. Rev. Neurosci. **14**, 531–570
11. Goodman, C.S. and Shatz, C.J. (1993) Cell **72**(Suppl.), 77–98
12. Tomasiewicz, H., Ono, K., Yee, D., et al. (1993) Neuron **11**, 1163–1174
13. Erickson, H.P. (1993) J. Biol. Chem. **120**, 1079–1081
14. Saga, Y., Yagi, T., Ikawa, Y., Sakakura, T. and Aizawa, S. (1992) Genes Dev. **6**, 1821–1831
15. Elkins, T., Zinn, K., McAllister, L., Hoffmann, F.M. and Goodman, C.S. (1990) Cell **60**, 565–575
16. Schachner, M. (1993) Curr. Opin. Cell Biol. **5**, 786–790

17. O'Leary, D.D., Heffner, C.D., Kutka, L., et al. (1991) Development 2(Suppl.), 123–130
18. Gundersen, R.W. and Barrett, J.N. (1980) J. Cell Biol. **87**, 546–554
19. Heffner, C.D., Lumsden, A.G.S. and O'Leary, D.D.M. Science **247**, 217–220
20. Serafini, T., Kennedy, T.E., Galko, M.J., et al. (1994) Cell **78**, 409–424
21. Kennedy, T.E., Serafini, T., de la Torre, J.R. and Tessier-Lavigne, M. (1994) Cell **78**, 425–435
22. Messersmith, E.K., Leonardo, E.D., Shatz, C.J., et al. (1995) Neuron **14**, 949–959
23. Colamarino, S.A. and Tessier-Lavigne, M. (1995) Annu. Rev. Neurosci. **18**, 497–529
24. Davenport, R.W., Dou, P., Rehder, V. and Kater, S.B. (1993) Nature (London) **361**, 721–724
25. Mattson, M.P., Dou, P. and Kater, S.B. (1988) J. Neurosci. **8**, 2087–2210
26. Zheng, J.Q., Felder, M., Connor, J.A. and Poo, M.M. (1994) Nature (London) **368**, 140–144
27. Schwab, M.E., Kapfhammer, J.P. and Bandtlow, C.E. (1993) Annu. Rev. Neurosci. **16**, 565–595
28. Oakley, R.A. and Tosney, K.W. (1993) J. Neurosci. **13**, 3773–3792
29. Tosney, K.W. and Landmesser, L.T. (1984) J. Neurosci. **4**, 2518–2527
30. Keynes, R.J. and Stern, C.D. (1984) Nature (London) **310**, 786–789
31. Keynes, R.J. and Stern, C.D. (1988) The Making of the Nervous System (Parnavalas, J.G., Stern, C.D. and Stirling, R.V., eds.), p. 84–100, Oxford University Press, Oxford
32. Davies, J.A., Cook, G.M.W., Stern, C.D. and Keynes, R.J. (1990) Neuron **2**, 11–20
33. Holt, C.E. and Harris, W.A. (1993) J. Neurobiol. **24**, 1400–1422
34. Walter, J., Henke-Fahle, S. and Bonhoeffer, F. (1987) Development **101**, 909–913
35. Walter, J., Kern-Veits, B., Huf, J., Stolze, B. and Bonhoeffer, F. (1987) Development **101**, 685–696
36. Baier, H. and Bonhoeffer, F. (1992) The Nerve Growth Cone (Letourneau, P.C., Kater, S.B. and Macagno, E.R., eds.), pp. 195–208, Raven Press, New York
36a. Drescher, U., Kremoser, C., Handwerker, C., et al. (1995) Cell **82**, 359–370
37. Kapfhammer, J.P., Grunewald, B.E. and Raper, J.A. (1986) J. Neurosci. **6**, 2527–2534
38. Kapfhammer, J.P. and Raper, J.A. (1987) J. Neurosci. **7**, 201–212
39. Kapfhammer, J.P. and Raper, J.A. (1987) J. Neurosci. **7**, 1595–1600
40. Raper, J.A. and Grunewald, B.E. (1990) Exp. Neurol. **109**, 70–74
41. Moorman, S.J. and Hume, R.I. (1990) J. Neurosci. **10**, 3158–3163
42. Luo, Y., Raible, D. and Raper, J.A. (1993) Cell **75**, 217–227
43. Kolodkin, A.L., Matthes, D.J. and Goodman, C.S. (1993) Cell **75**, 1389–1399
44. Kolodkin, A.L., Matthes, D.J., O'Connor, T.P., et al. (1992) Neuron **9**, 831–845
45. Schwab, M.E. and Thoenen, H. (1985) J. Neurosci. **5**, 2415–2423
46. Schwab, M.E. and Caroni, P. (1988) J. Neurosci. **8**, 2381–2393
47. Caroni, P. and Schwab, M.E. (1988) J. Cell Biol. **106**, 1281–1288
48. Caroni, P. and Schwab, M.E. (1988) Neuron **1**, 85–96
49. Schnell, L. and Schwab, M.E. (1993) Eur. J. Neurosci. **5**, 1156–1171
50. Cadelli, D.S., Bandtlow, C.E. and Schwab, M.E. (1992) Exp. Neurol. **115**, 189–192
51. Schnell, L. and Schwab, M.E. (1990) Nature (London) **343**, 269–272
52. Weibel, D., Cadelli, D. and Schwab, M.E. (1994) Brain Res. **642**, 259–266
53. Schnell, L., Schneider, R., Kolbeck, R., Barde, Y.-A. and Schwab, M.E. (1994) Nature (London) **367**, 170–173
54. Savio, T. and Schwab, M.E. (1989) J. Neurosci. **9**, 1126–1133
55. Kapfhammer, J.P. and Schwab, M.E. (1994) Eur. J. Neurosci. **6**, 403–411
56. Kapfhammer, J.P. and Schwab, M.E. (1994) J. Comp. Neurol. **340**, 194–206
57. Kapfhammer, J.P., Schwab, M.E. and Schneider, G.E. (1992) J. Neurosci. **12**, 2112–2119
58. Schwab, M.E. and Schnell, L. (1991) J. Neurosci. **11**, 709–722
59. Bandtlow, C., Zachleder, T. and Schwab, M.E. (1990) J. Neurosci. **10**, 3837–3848
60. Al-Mohanna, F.A., Cave, J. and Bolsover, S.R. (1992) Dev. Brain Res. **287**, 290–294
61. Mills, L.R. and Kater, S.B. (1990) Neuron **2**, 149–163
62. Anglister, L., Farber, I., Shahar, A. and Grinvald, A. (1982) Dev. Biol. **94**, 351–365
63. Fields, R.D., Guthrie, P.B., Russell, J.T., et al. (1993) J. Neurobiol. **24**, 1080–1098
64. Davenport, R.W. and McCaig, C.D. (1993) J. Neurobiol. **24**, 89–100
65. Mattson, M.P., Taylor-Hunt, A. and Kater, S.B. (1988) J. Neurosci. **8**, 1704–1711
66. Cohan, C.S., Connor, J.A. and Kater, S.B. (1987) J. Neurosci. **7**, 3588–3599
67. Mattson, M.P., Murrain, M., Guthrie, P.B. and Kater, S.B. (1989) J. Neurosci. **9**, 3728–3740
68. Mattson, M.P. and Kater, S.B. (1987) J. Neurosci. **7**, 4034–4043

69. Lipton, S.A. and Kater, S.B. (1989) Trends Neurosci. **12**, 265–270
70. Kater, S.B. and Mills, L.R. (1991) J. Neurosci. **11**, 891–899
71. Ivins, J.K., Raper, J.A. and Pittman, R.N. (1991) J. Neurosci. **11**, 1597–1608
72. Hess, D.T., Patterson, S.P., Smith, D.S. and Skene, J.H. (1993) Nature (London) **366**, 562–565
73. Forscher, P. and Smith, S.J. (1988) J. Cell Biol. **107**, 1505–1516
74. Bandtlow, C.E., Schmidt, M.F., Hassinger, T.D., Schwab, M.E. and Kater, S.B. (1993) Science **259**, 80–83
75. Haydon, P.G., McCobb, D. and Kater, S.B. (1984) Science **226**, 561–564
76. Rager, G.H. (1980) Development of the Retinotectal Projection in the Chicken, Springer Verlag, Berlin, Heidelberg, New York
77. Gierer, A. (1987) Proc. R. Soc. Lond. **218**, 77–93
78. Bonhoeffer, F. and Huf, J. (1980) Nature (London) **288**, 162–164
79. Walter, J., Allsopp, T.E. and Bonhoeffer, F. (1990) Trends Neurosci. **13**, 447–452
80. von Boxberg, Y., Deiss, S. and Schwarz, U. (1993) Neuron **10**, 345–357
81. Cox, E.C., Müller, B. and Bonhoeffer, F. (1990) Neuron **2**, 31–37
82. Zocchi, E. et al. (1993) Biochem. Biophys. Res. Commun. **196**, 1459–1465
83. Goldberg, D.J. and Burmeister, D.W. (1992) J. Neurosci. **12**, 4800–4807
84. Igarashi, M., Strittmatter, S.M., Vartanian, T. and Fishman, M.C. (1993) Science **259**, 77–79
85. Galione, A. et al. (1993) Nature (London) **365**, 456–459
86. White, A.M., Watson, S.P. and Galione, A. (1993) FEBS Lett. **318**, 259–263
87. Kim, H., Jacobson, E.L. and Jacobson, M.K. (1993) Science **261**, 1330–1333
88. Lankford, K.L. and Letourneau, P.C. (1989) J. Cell Biol. **109**, 1229–1243
89. Tanaka, J., Kira, M. and Sobue, K. (1993) Brain Res. Dev. Brain Res. **76**, 268–271
90. Lin, C.H., Thompson, C.A. and Forscher, P. (1994) Curr. Opin. Neurobiol. **4**, 640–647
91. Lin, C.H. and Forscher, P. (1993) J. Cell Biol. **121**, 1369–1383
92. Sobue, K. (1993) Neurosci. Res. **18**, 91–102
93. Garyantes, T.K. and Regehr, W.G. (1993) J. Neurosci. **12**, 96–103
94. Fields, R.D. and Nelson, P.G. (1994) J. Neurobiol. **25**, 281–293

8

Developmental aspects of axon growth and its inhibition in the vertebrate nervous system

Derryck Shewan* and James Cohen

Department of Development Neurobiology, UMDS, Guy's Campus,
London Bridge, London SE1 9RT, U.K.

Introduction

It has long been appreciated that, while axon regeneration in the adult mammalian peripheral nervous system (PNS) can often be successful, with a good chance of functional recovery, injured axons within the central nervous system (CNS) do not regrow and innervate their original targets. Inspired by the pioneering work of Tello [1], Aguayo's group circumvented inhibition of repair of damaged mammalian CNS axons [2,3] by bridging the proximal and distal stumps of axotimized CNS nerves with a piece of peripheral nerve. These experiments demonstrated the regeneration of axons in both directions through the peripheral nerve graft, only to be inhibited on contacting the CNS environment at the distal ends of the graft. Thus, Schwann cells, the glial cells of the PNS, combined with the abundant extracellular matrix (ECM) in peripheral nerves, provide a favourable environment that fosters regeneration after injury in the PNS and, when transplanted, can do so in the injured CNS. This provided convincing evidence that adult CNS neurons are not inherently unable to re-extend their axons, but rather the CNS glial environment is inhospitable to axon outgrowth, and prompted what has become an extensive search for the inhibitory molecule(s) responsible for this phenomenon. Moreover, both the macroglial cell types of the mature CNS, oligodendrocytes and astrocytes, are likely to effect this inhibition, since nerve regeneration has been shown to fail in both grey and white matter of the CNS [4].

The ability of adult CNS neurons to regenerate, however, even in the presence of sciatic nerve grafts, varies over a wide range [5]. Moreover, as neurons age, new constraints are likely to be imposed on the regenerative response, due to changes in neuronal phenotype that accompany maturation [6–8]. Thus, transplantation studies carried out by Björklund's group [9,10] have shown that early embryonic human forebrain neuroblasts can extend long axons within the adult rat

* To whom correspondence should be addressed.

brain, where axon regeneration is normally inhibited. Moreover, these immature axons often successfully innervate appropriate target areas of the CNS region into which the neuroblasts are transplanted. A similar phenomenon was reported more recently by Li *et al.* [11], who showed that early post-natal neurons of the rat entorhino-hippocampal pathway regenerate their axons in slice culture, but neurons a week later in development do not. Thus, the axons of immature neurons appear not to recognize molecules that inhibit the growth of their more mature counterparts. Such studies of axon growth mechanisms in early development may provide ideas as to how axons extend towards their targets, and may therefore point to differences in the injured adult mammalian CNS that are responsible for the failure of axon regeneration.

This chapter reviews the best-characterized molecules that have been shown to inhibit axon outgrowth both *in vivo* and *in vitro*. Arguably, for accurate axon pathfinding, the inhibition of axon growth is as important as its promotion; since only by defining 'no entry' zones can an axon follow a defined pathway, i.e. by being discouraged from straying from the appropriate route. Furthermore, on reaching its target, an axon must stop growing and be prevented from branching inappropriately. The list compiled in Table 1 represents inhibitory molecules that have been identified in a variety of loci (including within the ECM, on the surfaces of neurons and CNS macroglial cells, and in CNS myelin), as well as secreted factors that may function in solution or by binding specifically to components of the ECM. After discussing the putative inhibitory roles of these molecules in axon growth and regeneration, we present recent data from this laboratory using an *in vitro* technique that may allow clarification of the mechanisms of axon growth and its inhibition, and which may help identify developmental neuronal changes responsible for the lack of regeneration of injured adult CNS axons.

Axon growth-inhibitory/repulsive molecules

There are presently considered to be two major, functionally distinct mechanisms of preventing axon growth. Some molecules are thought to deflect the outgrowth of an axon by providing a substratum less conducive to growth than an adjacent substratum, so that the axon may be repelled without arresting or retracting. Other molecules may actively induce the collapse of the motile sensory tip of the growing axon, the growth cone, resulting in a temporary, at least, arrest of growth, and often a considerable retraction of the axon. This latter phenomenon is now considered by many to represent classical axon growth inhibition. However, the majority of the molecules described below have yet to be categorized as inducing one mode of inhibition or the other, and indeed many have been reported to cause both. Thus, except where strong evidence suggests otherwise, no attempt has been made to distinguish between the modes of inhibition they mediate. It should be noted that

Table 1   Axon growth-inhibitory molecules

Factor	Size	Cell type	Mode of attachment	Location or source	Reference
Tenascin	200/220 kDa	Astrocytes, Schwann cells	ECM	DREZ, scar tissue, ECM	[31]
J1-160/180	160/180 kDa	Oligos?	ECM/secreted	Myelin	[33]
S-laminin	200 kDa	Muscle	ECM	Neuromusc. junction	[36]
CSPG					
KSPG	Variable	(Reactive) Astrocytes	ECM	Roofplate/scar tissue	[16,23]
Collapsin/semaphorins	~ 100 kDa	Neurons, muscle cells	Secreted/transmembrane	Neurons, muscle and epithelia	[39,40] [41]
Post. tectal glycoprotein	25/33 kDa	?	GPI-linked	Optic tectum	[48,50]
PNA-binding glycoprotein	55 kDa	Sclerotome	Membrane bound?	Posterior somite	[55]
Thy-1	18 kDa (Ig domain)	Neurons, glia	GPI-linked	PNS and CNS	[58]
Connectin	72 kDa	Muscle cells, motor neurons, glial cells	GPI-linked	Drosophila muscle and nervous system	[57]
NI-35/250	35/250 kDa	Oligos	Transmembranous	CNS myelin	[62]
MAG	~ 100 kDa (Ig domains)	Oligos	Transmembranous	CNS and PNS myelin	[63,64]
Netrin-1/2	75/78 kDa (unc-6 homologue)	Ventral floorplate	Secreted	Ventral spinal cord	[68,69]
Unknown soluble factor	(Netrin?)	Septal neurons	Secreted	Septum and ventral spinal cord	[67]
Nitric oxide	Gaseous	Many	Soluble	Widespread	[72]

Abbreviations: DREZ, dorsal root entry zone; oligos, oligodendrocytes; post., posterior.

few of the axonal cell-surface receptors that transduce the axon growth-inhibitory signals have been identified.

ECM molecules

Astrocytes, the class of macroglia found in both grey and white matter, express molecules in the injured CNS that have been shown to inhibit axon extension *in vivo* and *in vitro*. On maturation, mammalian astrocytes may undergo molecular and functional changes that are responsible for axon growth inhibition [12,13], and, perhaps, for the inhibition of regeneration by injured sensory neurons of the adult rat, preventing re-innervation of the spinal cord by sensory afferents [14]. For example, astrocytes express and secrete various proteoglycans (PGs) [15]. Chondroitin sulphate PG (CSPG) may contribute to the appropriate orientation of retinal ganglion cell (RGC) axons within the retina towards the optic nerve by repelling them from the periphery of the retina ([16,17]; see also Chapter 6), and inhibits neurite outgrowth by chick RGCs *in vitro* [18]. CSPG is also transiently up-regulated in the mature CNS after injury [19]. Moreover, neurocan, a developmentally regulated CSPG of rat brain, binds to the cell adhesion molecules (CAMs), neural cell adhesion molecule (NCAM) and L1, which normally promote neurite outgrowth, and causes growth inhibition, suggesting that PGs can modulate the function of other axon growth-regulating molecules [20]. In contrast, however, further evidence suggests that CSPG in the developing mouse does not present a barrier to outgrowing thalamocortical neurons [21], and that a CNS-specific CSPG can promote neurite outgrowth by rat embryonic day (E) 14 mesencephalic and E18 hippocampal neurons *in vitro* [22].

Other subclasses of PGs may also exert inhibitory influences during development; for example keratan sulphate PG, which may underlie the prevention of crossing of commissural and dorsal cord axons at the dorsal midline of the spinal cord [23]. It is also expressed in midline structures of the developing chick CNS [24], but not in animals whose adult CNS axons are able to regenerate, such as the goldfish, frog and leech [25]. More recently, a novel PG of molecular mass 160–220 kDa was isolated from CNS scar tissue, and inhibited neurite outgrowth by embryonic rat dorsal root ganglion (DRG) septal and hippocampal neurons *in vitro* when presented as the substratum or in soluble form [26]. Thus, the expression of such molecules by astrocytes may explain why axon regeneration fails within scar tissue of grey matter tracts as well as in white matter areas [4], and why neurite outgrowth on optic nerve tissue devoid of oligodendrocytes and CNS myelin is inhibited [27]. The mechanisms by which PGs may alter the growth and orientation of axons are dealt with elsewhere in this book (Chapter 6).

Matrix molecules other than PGs may also both promote and inhibit axon outgrowth in different contexts, possibly by binding to other growth-regulatory molecules. Tenascin, a hexameric matrix molecule expressed by astrocytes [18], may have both growth-repelling and growth-supporting roles [28–30]. It is transiently expressed in developing mouse optic nerve, is not permissive for neurite outgrowth *in vitro*, and thus may contribute to the inhibitory environment encountered by injured CNS axons [31]. However, it is also up-regulated after injury of mature peripheral nerve [32], which can regenerate successfully, and thus its function remains unclear.

Janusin, a tenascin-like molecule with 160 kDa (monomeric and dimeric) and 180 kDa (trimeric) isoforms, is also reported to have conflicting effects on axon outgrowth. It is expressed *in vitro* by oligodendrocytes and type-2 astrocytes from the post-natal mouse optic nerve and spinal cord [33], and in late embryonic rat hippocampus and cerebral cortex [34]. Although E6 chick retinal and E7/8 DRG neurites are predominantly repelled on encountering janusin *in vitro* [35], the relatively late expression of janusin suggests that it may be involved in regulating collateral sprouting, rather than in initial pathfinding.

Other, less well characterized matrix molecules may prove to play prominent roles in the inhibition of axon growth. For example, Porter *et al.* [36] propose that a tripeptide sequence in S-laminin, a homologue of the B1 laminin chain, may selectively inhibit motor neuron axon outgrowth in development. In this way it may act as a 'stop' signal to axons innervating skeletal muscle, where S-laminin expression is concentrated at the synaptic basal lamina. Thus, matrix molecules may possess significant inhibitory functions on target contact that result in the cessation of cytoskeletal polymerization and the stabilization of axons.

Molecules with collapsing activity

The establishment of axonal pathways may involve the prefiguring of inhibitory boundaries which deter axons from straying from their correct course and innervating inappropriate targets. A common feature of such contact-inhibited growth is to induce filopodial retraction and disruption of lamellipodial structure, designated growth cone collapse [5]. Raper was the first to characterize a growth cone collapsing activity, based on initial observations by Bray *et al.* [37], demonstrating paralysis of sympathetic growth cones on contact with CNS (retinal) neurites [38]. A 100 kDa glycoprotein termed collapsin, which is responsible for this activity, has been purified from chick brain, and the gene encoding it has been cloned [39]. When transfected into COS-7 cells, co-cultured embryonic chick DRG neurons are inhibited from extending processes. Moreover, it exhibits strong homology with fasciclin IV, which is a transmembrane protein expressed by some grasshopper axons of the CNS and epithelial cells of the limb bud [40]. Antibody-blocking of fasciclin IV results in abnormal pioneer axon extension in the grasshopper limb bud. Collapsin and fasciclin IV are two members of an expanding gene family of transmembrane and secreted molecules, termed semaphorins, that may play important roles in growth cone guidance during initial axon outgrowth. Members of this family exhibit a homology that is conserved between the nervous systems of vertebrates and invertebrates [40,41]. Indeed, five new members of the semaphorin family have recently been identified in the mouse, and one (semaphorin D) is highly homologous to collapsin [42]. Furthermore, it appears that the function of such axon growth-inhibitory molecules may be highly specific, since murine semaphorin III appears to inhibit outgrowth by only a subset of embryonic rodent sensory neurons [43].

Growth cone collapse assays will doubtless lead to the identification of novel molecules in this category, and have already demonstrated, for example, such a mode of action for the inhibitory myelin proteins discussed below.

The retinocollicular pathway

The role of putative growth cone collapse-inducing proteins, in tandem with axon-repulsive cues, in directing axon outgrowth and target innervation is possibly best exemplified in the retinocollicular pathway, the primary visual projection from the retina to the brain. Thus, the mammalian retinocollicular pathway makes an excellent subject for studying axon growth mechanisms, since different populations of optic axons must respond to different growth cues, particularly at the chiasm and within the colliculus. Furthermore, axon regeneration within the optic nerve has been extensively studied due to its relative simplicity and its accessibility for surgery.

The specificity and accuracy of the topography of innervation of the tectum by optic axons was recognized by Sperry [44], who performed classical experiments in amphibians, involving optic nerve regeneration, that led to our present understanding of the complexity of the mechanisms of axon outgrowth and target recognition. Thus, axons regenerating from the denervated retina of the frog are dependent on the orientation on the retina for topographically correct re-innervation in the optic lobe. If the retina is rotated through 180 ° before nerve section, the orientation of visuomotor responses after recovery is accordingly reversed. Most impressive, however, was the projection of lesioned axons of the optic nerve to establish functional connections in their original loci of the optic lobe. Thus, there must exist highly specific molecular cues that support and guide axon outgrowth to specifically localized areas of the brain.

In the retinotectal pathway, RGCs project axons along the optic nerve into the optic tract via the optic chiasm, and then innervate the tectum. At the chiasm in mammals, a subpopulation of RGC (ventrotemporal) axons do not cross over the midline, as other RGC (nasal) axons do to innervate the contralateral colliculus, but project ipsilaterally. One explanation for the different axonal behaviour at the optic chiasm in mammals is that an intrinsic scaffolding exists in this region that permits the outgrowth of retinal axons [45]. An inherent population of axons at the optic chiasm of the developing mouse express L1, a member of the immunoglobulin superfamily of CAMs most of which promote axon outgrowth; L1 expression extends along the optic tracts toward the ventral diencephalon, while CD44, a CAM first described in the immune system that inhibits retinal neurite outgrowth in vitro, is expressed at the midline of the chiasm, and thus may constitute a selective barrier for ipsilaterally projecting axons [45]. Although no attempt was made to test the different subpopulations of RGCs that exhibit opposing behaviour at the chiasm, it may be that only ventrotemporal RGC axons recognize CD44 as an inhibitory ligand. The idea that inherent axonal scaffolds may guide projecting axons is not an uncontested one, however. Cornel and Holt [46] have demonstrated that optic axons from grafted eye primordia of *Xenopus laevis* can navigate appropriately to innervate the tectum when the formation of inherent axonal scaffolding, the tract of the post-optic commissure along which retinal axons normally fasciculate, is delayed by prior hydroxyurea treatment. Thus, it is likely that other local cues must exist to guide retinal axons to their targets. Elegant *in vitro* experiments by Wizenmann *et al.* [47], using embryonic rat retinal explants grown on membrane fractions of colliculus and chiasm, show that the ventrotemporal axons do not extend axons on chiasm membranes, but only when plating E17/18 explants on E14/15 chiasm membranes. Neurites of RGCs originating from other retinal areas showed no inhibition on chiasm preparations of any age. Moreover, ventrotemporal RGC neurites were not inhibited on membranes of other CNS areas known to be inhibitory to other neural populations. Thus the inhibitory activity that prevents ventrotemporal RGC axons from crossing the chiasm, a step fundamental in achieving binocular vision and depth of field in mammals, is highly specific and age-dependent, and based on a selective growth cone inhibitory mechanism.

As discussed above, RGC axons innervating the tectum are topographically mapped to specific loci. The molecular basis underlying this phenomenon is unknown, but recent studies suggest inhibitory molecules may play a part. Thus, Drescher *et al.* [48] have identified a 25 kDa glycosylphosphatidylinositol (GPI)-anchored glycoprotein, repulsive axon guidance signal (RAGS), expressed in the posterior tectum of the developing chick that may be important in patterning the innervation of the tectum by optic axons. RAGS bears sequence similarity to ligands of the Eph receptor tyrosine kinase family [49], but shows no specificity for inducing collapse of growth cones of temporal, rather than nasal, RGC axons. This suggests a general inhibitory role, and not one of specific axon guidance by growth inhibition as previously reported for a 33 kDa GPI-anchored molecule, also temperospatially expressed in the posterior tectum [50,51]. While RAGS appears to have a general inhibitory effect on retinal axons, Drescher *et al.* [48] suggest that it may interact with the 33 kDa molecule that specifically inhibits temporal axons, thus contributing to a specific guidance phenomenon by preventing temporal retinal axons from extending into the posterior region of the tectum [52]. Moreover, deafferented adult rat superior colliculus may re-express the 33 kDa inhibitory glycoprotein, and both embryonic and adult RGCs can respond to the re-expressed activity [53].

Evidence for the existence of a similar, GPI-anchored molecule in mammals has recently been reported by Roskies and O'Leary [54], who suggest that the inhibitory activity in the caudal superior colliculus of developing rats may selectively induce collateral branching and arborization of appropriate retinal axons in the topographically correct rostral, but not caudal, region of the colliculus, thus

establishing topographically ordered synaptic connections. Thus, a mammalian inhibitory glycoprotein expressed in equivalent collicular areas to the 33 kDa molecule described by Bonhoeffer in the chick tectum may exert its effects on projecting temporal RGC axons by a subtly different mechanism.

The retinocollicular pathway, then, provides excellent examples of how axon growth-inhibitory molecules may be involved in pathfinding by developing and regenerating axons. Further classes of growth inhibitors are considered below.

Membrane-bound glycoproteins

This category may include a peanut agglutinin (PNA)-binding glycoprotein restricted to posterior half somites that induces growth cone collapse *in vitro*, and is thought to repel both migrating neural crest cells and motor axons *in vivo*, thus confining them to the anterior half somites, a critical step for spinal segmentation [55]. When a PNA-binding glycoprotein fraction, isolated from chick somites by affinity chromatography, was presented as a substratum for embryonic chick DRG explants, neurite outgrowth was substantially reduced compared with control cultures. Furthermore, in explant cultures on laminin in the presence of liposomes incorporated with sclerotome-derived proteins, the incidence of growth cone collapse increased to about three times that in control cultures [56].

A further candidate, in *Drosophila*, that may exert inhibitory influences on developing motor axons is connectin. This is a cell-surface, homophilic-binding protein with a putative GPI-linked attachment that is expressed on a subset of embryonic muscles and the motor axons that innervate them, as well as on some associated glial cells [57]. Ectopic expression of connectin on inappropriate muscles results in aberrant innervations by motor axons, and has revealed that this molecule may have both axon growth-promoting and inhibitory influences on different subsets of neurons, again suggesting that axon pathfinding cues can be highly specific.

Thy-1 is a small GPI-anchored glycoprotein of the immunoglobulin superfamily that has a molecular mass of 25 kDa and is expressed in several tissues, including the nervous system. It is expressed by many classes of neurons after target contact in development has been established. Anti-Thy-1 antibody-blocking of a Thy-1-expressing neural cell line reverses the inhibition of neurite extension observed on astrocytes [58]. Similarly, neurite outgrowth by neonatal rat superior cervical ganglion (SCG) neurons exhibits a striking increase when cultured in the presence of anti-Thy-1 antibodies [59]. Further evidence for Thy-1 homophilic binding is provided by immunoblots of neonatal rat SCG neurons and pheochromocytoma cells (PC12 cells), which demonstrated bands of 45 kDa and 150 kDa, suggesting multimeric forms of Thy-1 may exist *in vivo* [60]. Thus, Thy-1, the smallest member of the immunoglobulin superfamily, appears to potently inhibit neurite outgrowth, and may be partly responsible for the prevention of axon growth at a late stage of development and, perhaps, after injury.

CNS myelin proteins

In mammals, two oligodendrocyte cell-surface glycoproteins (NI-35/250) expressed in CNS myelin have been extensively studied in relation to their axon growth-inhibitory influence on regenerating neurons [61,62]. These proteins are discussed by Bandtlow in Chapter 7 of this volume, and thus are not considered further here.

Myelin-associated glycoprotein (MAG), a member of the immuno-globulin superfamily that is expressed in both the CNS and, to a lesser extent, the PNS, has more recently been attributed a growth-inhibitory role for axon outgrowth by some classes of neurons. It is permissive for neurite outgrowth by neonatal, but not adult, rat DRG neurons, and also inhibits outgrowth by neonatal cerebellar neurons when expressed on the surface of fibroblasts in vitro [63]. Furthermore, neurite extension by the neuroblastoma cell line NG108-15 is blocked when cultured on recombinant MAG [64]. In the latter case, immuno-depletion of MAG resulted in restoration of neurite outgrowth to about 63% of the control level. However, the relatively late expression of myelin proteins in development, coupled with the evidence that axon growth inhibition occurs in the CNS before myelination [27], suggests they do not play a prominent role in initial axon pathfinding in development. They may, however, be important in regulating collateral branching once the axonal pathway is established, with the consequence of blocking regeneration after injury to mature CNS axons. However, they are unlikely to be solely responsible for the lack of axon regrowth in the injured adult CNS, since regeneration also fails within myelin-free grey matter [4,65].

Soluble factors

In the rat, a specific age-dependent chemorepulsive factor in early embryonic (E14–15), but not late embryonic (E18), ventral spinal cord selectively inhibits neurite outgrowth from DRG neurons, but not dorsal cord [66]. Chemorepulsion may also prevent inappropriate pathway formation by developing rat lateral olfactory tract axons [67]. These unidentified factors may be netrin or netrin-like molecules. The netrins are recently identified molecules with primary structures consistent with secreted proteins, and promote and guide the growth of developing commissural axons in vitro [68,69]. However, netrin-1-secreting COS cells implanted into collagen gels inhibit neurite outgrowth by trochlear motor neurons [70]. Furthermore, the UNC-6 protein in Caenorhabditis elegans, which is chemotactic for developing commissural axons and homologous to the netrins, may also inhibit migration of mesodermal cells [71]. Thus, such soluble factors may be highly specific in function, and may serve dual purposes in development, depending on cell type.

Nitric oxide (NO), which is increasingly being investigated in a variety of cell types and physiological systems, may also play a role in nerve development and regeneration. Thus, regenerating adult rat DRG neurites on a laminin substratum, when subjected to NO, collapse their growth cones and frequently retract [72]. The physiological significance of this observation remains to be determined. Two major neurotransmitters of the CNS, glutamate and 5-hydroxytryptamine (serotonin), may also inhibit axon outgrowth in some conditions. Local application of the excitatory amino acid, glutamate, induces dose-dependent dendritic growth cone collapse of late embryonic (E18) rat hippocampal neurons grown in culture on a polylysine substratum, a response that may be mediated by an increased intracellular Ca^{2+} concentration ($[Ca^{2+}]_i$) [73], while 5-hydroxytryptamine induces a neuron-selective, dose-dependent inhibition of adult *Helisoma* ganglion cells in culture [74].

Thus, a variety of molecules have now been identified that may function as inhibitors of axon outgrowth, both in development and regeneration. They encompass a diverse group of molecules, including ECM components, (tenascin, janusin, PGs); secreted proteins (collapsin, netrins and some invertebrate semaphorins); and transmembrane and GPI-anchored glycoproteins (CNS myelin proteins, fasciclin IV/semaphorin I, 33 kDa posterior tectum glycoprotein, Thy-1). It is unlikely that the above list is exhaustive, however, and by analogy with the range of axon growth-promoting molecules so far identified, many more classes of repulsive molecules probably remain to be discovered. From the wealth of evidence involving candidate molecules mediating axon growth and inhibition, it is clear that extending neural processes must negotiate complex environments that exert specific, often age-dependent influences on different neuronal classes both in development and regeneration. Thus, outgrowing axons most likely require a balance of supportive and repulsive guidance cues to reach their targets, with inhibitory molecules possibly excluding axons from inappropriate areas and, thus, confining them to specific pathways.

A further aspect of these studies is that axons must possess the appropriate receptors to recognize those ligands that influence axon growth. This is the most likely explanation for the impressive axon outgrowth achieved when early human fetal neuroblasts are transplanted into the mature rat CNS [9,10], since it is possible that, at early embryonic ages, neurons lack receptors for putative CNS inhibitory ligands. The same may be true for transplanted adult mouse subventricular zone cells that can migrate over long distances in the mature CNS and differentiate into neurons of the olfactory bulb [75]. Thus, it is insufficient to solely consider the glial environment of the CNS; the intrinsic properties of developing and regenerating neurons, as well as the second messenger pathways initiated by binding their cognate receptors on axons, need also to be better understood.

Bridging the gap between *in vitro* and *in vivo* experimental approaches

The number and variety of putative axon growth-inhibitory molecules is extensive (Table 1), and indeed this list continues to grow. However, the precise mechanisms by which axons grow, and are prevented from doing so in the injured mammalian CNS, remain elusive. In this regard, there are unavoidable drawbacks in both *in vitro* and *in vivo* techniques designed to gain insight into these mechanisms. Much of the *in vitro* work carried out to test neurite outgrowth responses oversimplifies the neuronal environment, making interpretation of the results difficult, with respect to the *in vivo* situation. Nevertheless, *in vitro* tests are necessary to simplify the complex nature of the *in vivo* environment, since (i) the wide variety of cell types that may influence axon growth, (ii) the possibility of unknown soluble factors being secreted by these different cell types, as well as (iii) the infliction of injury (thus effecting injury-induced responses within the manipulated tissues) combine to obscure the actual mechanisms underlying axon growth inhibition. Thus, it remains unclear as to what cell types inhibit or repel axons in development and after injury, and what soluble trophic/tropic factors are required to promote (i) neuronal survival, (ii) directed growth of axons to their targets, and (iii) synapse formation with those targets. All of these factors are necessary to develop a functional nervous system, and, in the adult, to maintain synaptic plasticity, as well as to foster regeneration of injured CNS axons in a clinical context.

In order to simplify the *in vivo* axonal environment, but maintain a measure of physiological integrity, we have adopted a cryoculture technique [76–78] in which thin sections of unfixed, frozen nerve tissue are employed as substrata for dissociated primary neurons. Thus, the ECM and cell-surface components of the nerve tissue remain largely intact, while the possible influences of soluble factors released by living cells are minimized [79]. This is exemplified in Fig. 1, which shows an adult rat DRG neuron (labelled with an antibody against the cytoskeletal component β-III tubulin; Fig. 1A), growing on an adult rat sciatic nerve cut in transverse section (labelled with an antiserum to laminin; Fig. 1B). The laminin-positive endoneuria of the underlying axon tubes are clearly visible (Fig. 1B), and by comparing the neurite outgrowth observed in Fig. 1(A), it is evident that many neurites appear to loop around these structures (arrows).

By using this physiologically appropriate technique, we can attempt to test the molecular mechanisms that promote or inhibit axon growth in developing and regenerating nerves. Moreover, we can compare and contrast the effect of ECM and cell-surface components on regenerating neurites without inducing an injury response, as *in vivo* manipulations inevitably do. In the data presented here, the optic nerve of the rat was chosen as a representative, yet fairly simple, CNS tissue. In the adult, the optic nerve largely consists of CNS macroglia, CNS myelin and RGC axons, while in the neonate there are no oligodendrocytes or myelin. As discussed above, mature CNS macroglial cells, and CNS myelin proteins, have been strongly implicated in the inhibition of axon growth in the injured mammalian CNS. By using optic nerves of immature and adult rats as substrata, we can directly test the influence of the presence or absence of oligodendrocytes and CNS myelin on neurite outgrowth. On these two substantially different CNS substrata we plated neurons of various developmental ages. RGCs were chosen as clearly being the most appropriate for regeneration within an optic nerve environment. DRG neurons were also used since, in development, they extend processes within both the PNS and CNS, and thus normally encounter both astrocytes and oligodendrocytes. Furthermore, by plating neurons of various ages, while keeping all other conditions constant, we can test the hypothesis that axon growth and its inhibition may be dependent on maturational changes within the neurons themselves, as suggested by the work of Björklund's and Raisman's groups [9–11].

Fig. 1. **Fluorescent photomicrographs of an adult rat DRG neuron labelled with an antibody against β-III tubulin, a component of the cytoskeleton (A), extending neurites over a thin section of an adult rat sciatic nerve cut in transverse section, revealed with an antiserum against laminin (B)**

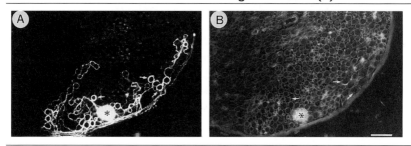

*Neurites are often seen looping around the laminin-positive endoneuria of the transversely and obliquely cut axon tubes in the underlying nerve section (corresponding to arrows in **A** and **B**). Asterisk indicates the neuronal soma. Scale bar = 50 μm.*

With these hypotheses in mind, we have shown that neurite outgrowth by neonatal rat RGCs is completely inhibited on both adult and neonatal rat optic nerve tissue sections, but is extensive on a pure substratum of the laminin homologue, merosin [27]. In contrast, E15–E16 RGCs do extend lengthy processes on both immature and mature optic nerve tissue sections [80]. Similarly, both adult and neonatal rat DRG neurons fail to extend neurites on the neonatal and adult optic nerve substrata, but E14–E15 DRG neurons grow extensively on both types of section [80]. Examples of this age-dependent switch in neurite outgrowth behaviour are shown in Fig. 2. The fact that neonatal RGCs labelled with an antiserum against the growth-associated protein GAP43 fail to grow on either immature (Figs. 2A and 2B) or adult (Figs. 2C and 2D) optic nerve substrata demonstrates that inhibitory molecules other than CNS myelin proteins exist in the immature CNS that influence axon growth and guidance, and are likely to be expressed by astrocytes or RGC axons themselves. Moreover, the dramatic, extensive neurite outgrowth observed in the case of E16 RGCs on both optic nerve substrata (Figs 2E–2H) concurs with previous work [8–11] in suggesting that maturation-dependent changes in neuronal phenotype, most likely an up-regulation of receptors for CNS inhibitors, occurs at a late stage of embryonic development.

Fig. 2.   **Fluorescent photomicrographs of double-immunolabelled**
                     cultures of neonatal (A–D) and early embryonic (E–H) rat
                     RGCs plated on sections of unmyelinated, immature (A,B;
                     E,F) and myelinated, adult (C,D; G,H) rat optic nerve

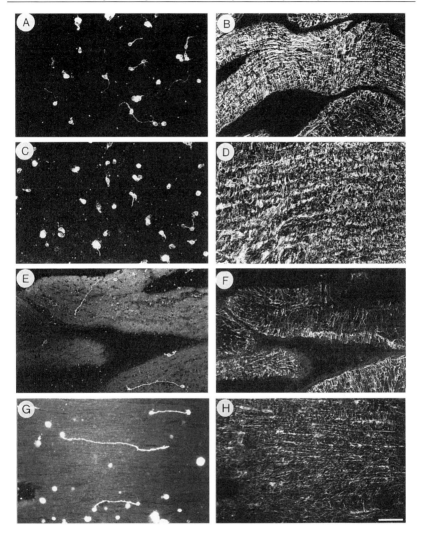

*GAP43-positive neonatal (P2) RGCs do not extend long neurites on either immature (A,B) or adult (C,D)
optic nerve, but some early embryonic (E15) RGCs do regenerate extensively on both immature (E,F) and
adult optic nerve (G,H). (A, C, E and G: GAP43; B, D, F and H: GFAP). Scale bar = 1 00 μm.*

Fig. 3.   **Histograms to represent the age-dependent growth response of rat RGC (A,B) and DRG neurons (C,D) on sections of unmyelinated, immature (A,C) and myelinated, mature (B,D) rat optic nerve**

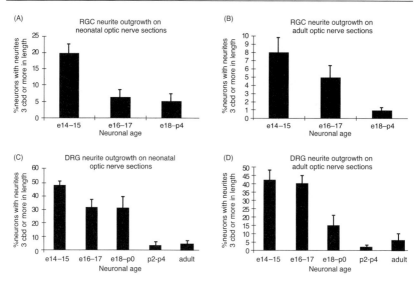

Data represent the percentage of GAP43-positive neurons that extend neurites of at least 3 cell body diameters (cbd) in length.

Our results are summarized in Fig. 3, which shows the influence of neuronal age on the ability to extend neurites over optic nerve cryosections *in vitro*. The age-dependent decline in neurite outgrowth by RGCs is evident on both neonatal and adult optic nerve (Figs. 3A and 3B), although it is more striking on immature optic nerve where almost 20% of early embryonic neurons extend long neurites. More impressive, however, is the regenerative response by early embryonic DRG neurons, almost 50% of which grow extensively on both immature and mature optic nerve (Figs. 3C and 3D). This response declines sharply to under 5% at a neonatal neuron age. Thus, DRG neurons, at least, exhibit an age-dependent, rapid and dramatic increase in neurite growth inhibition on CNS tissue substrata at E17–E18, the time when, *in vivo*, the axons of these neurons make contact with their peripheral and central targets [81]. To a lesser extent, RGC axons may also up-regulate receptors for CNS inhibitors at around the time they innervate the colliculus, which is their target tissue *in vivo*. The first axons arrive there at about E16 in the rat, and innervation continues into the post-natal period. This study, and those of Wictorin *et al.* [9,10] and Li *et al.* [11], emphasize that inhibitory molecules already expressed in the developing CNS may not be

Fig. 4. Schematic diagram representing the hypothesis that axon regeneration in the CNS fails due to the irreversible up-regulation of neuronal cell-surface receptors for CNS axon growth-inhibitory molecules already expressed by astrocytes in the immature nervous system

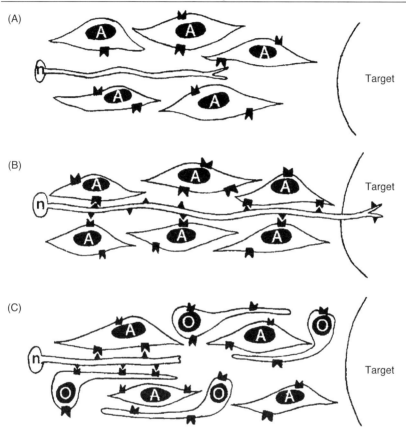

At early embryonic ages, axons can grow through the glial environment since they do not recognize CNS growth inhibitors (**A**). When target contact is established (**B**), receptors for CNS growth inhibitors are up-regulated to prevent inappropriate collateral branching. Subsequent injury to CNS axons is not followed by regeneration due to the persistence of the receptors to CNS inhibitors (**C**). Abbreviations: A, astrocyte; O, oligodendrocyte; n, neuron; ▲▲ , inhibitory ligand; ▲, neuronal cell-surface receptor for inhibitory ligand.

recognized by early developing neurons, either because the axonal receptors required to interact with the inhibitors are not yet expressed, or the receptors are as yet unable to transduce the inhibitory signal. Furthermore, the timing of the acquired neurite outgrowth inhibitory responses in late development closely coincides with the time when the majority of DRG axons and RGC axons contact their targets *in vivo*. Thus it seems likely that, as shown in Fig. 4, developing axons can grow through a potentially inhibitory environment to reach their targets, at which point they consequently up-regulate receptors for ECM/cell-surface inhibitory molecules thereby preventing further outgrowth or inappropriate collateral sprouting. It also seems likely that these receptors persist to adult stages, but are not down-regulated subsequent to nerve injury, thus preventing the regrowth of damaged CNS axons.

Future directions

While knowledge of putative axon growth-inhibitory molecules continues to grow, the neuronal receptors to which these inhibitory ligands bind remain to be discovered. This should now be a priority, since their identification should boost our knowledge of axon growth inhibition, and the search for means of fostering axon regeneration in the mammalian CNS. Of equal importance is the further clarification of neuronal second messenger systems associated with these putative receptors, that transduce external signals to a variety of loci within the neuron, thus eliciting an axon growth response. The complexities of these intracellular pathways continue to be unravelled, and are discussed elsewhere in this volume. Furthermore, the identification of candidate axon growth-inhibitory molecules is set to continue, adding members to groups and families of molecules conserved over a variety of species, such as the semaphorins and, perhaps, the netrins [41,42,68,69,82]. This evolutionary conservation may reflect the important role axon growth inhibition plays in axon guidance. Moreover the increasing number of putative inhibitory molecules may reflect the high specificity in axon–substrate interactions, which is necessary for the precise guidance of axons to their targets.

Some of the difficulties associated with *in vitro* and *in vivo* techniques designed to clarify axon growth mechanisms may be overcome by genetic manipulation, which can delete or overexpress single genes that encode molecules thought to play important roles in axon growth and its inhibition. To date, the outcome of such experiments has been relatively surprising, since the phenotype of mice manipulated in this way is often unchanged, or shows only minor differences. Examples include mice lacking NCAM [83] and tenascin [84]. The common explanation for such results is that the animal is capable of developing in a fairly normal fashion due to the existence of 'backup' molecules that can assume the role(s) of those that are 'knocked out', or the existence of other molecules that have similar functions. As technology improves, however, it is now possible to delete multiple genes, and thus elucidate molecules that influence axon growth together, rather than singularly, or which may assume the function of deleted molecules in abnormal development. More importantly, it is now possible to delete genes at strategic points of development, thus pinpointing the respective contributions of molecules implicated by *in vitro* studies in axon growth and its inhibition. In combination with *in vitro* techniques testing the regenerative abilities of neurons from gene-knockout animals, as well as the neurite outgrowth responses on their nerve tracts when employed as culture substrata, such animal models should provide further insight into the mechanisms of axon growth and its inhibition, and provide ideas as to how the failure of regeneration in the injured adult CNS may be circumvented.

We thank Kevin Fitzpatrick and Sarah Smith for excellent photographic assistance, and Professor Martin Berry for research facilities and surgical manipulation. The GAP43 antiserum was a gift from Graham Wilkin (Imperial College, London), and the laminin antiserum kindly given by Janet Winter (Sandoz Institute, Gower Street, London). The work shown in this chapter was carried out during the tenure of a studentship awarded (to D.S.) by the Anatomical Society for Great Britain and Ireland.

References

1. Tello, S. (1907) Trav. du Lab. de Rech. Biol. **5**, fasc. 3
2. David, S. and Aguayo, A.J. (1981) Science **214**, 931–933
3. Benfey, M. and Aguayo, A.J. (1982) Nature (London) **296**, 150–152
4. Maxwell, W.L., Follows, R., Ashurst, E. and Berry, M. (1990) Phil. Trans. R. Soc. London, [Biol.] **328**, 479–500
5. Fawcett, J.W. (1992) Trends Neurosci. **15**, 5–8
6. Cohen, J., Burne, J.F., Winter, J. and Bartlett, P.F. (1986) Nature (London) **322**, 465–467
7. Fawcett, J.W., Housden, E., Smith-Thomas, L. and Meyer, R.L. (1989) Dev. Biol. **135**, 449–458
8. Bedi, K.S., Winter, J., Berry, M. and Cohen, J. (1992) Eur. J. Neurosci. **4**, 193–200
9. Wictorin, K., Brundin, P., Gustavii, B., Lindvall, O. and Björklund, A. (1990) Nature (London) **347**, 556–558
10. Wictorin, K., Brundin, P., Sauer, H., Lindvall, O. and Björklund, A. (1992) J. Comp. Neurol. **323**, 475–494
11. Li, D., Field, P.M. and Raisman, G. (1995) Eur. J. Neurosci. **7**, 1164–1171
12. Smith, G.M., Rutishauser, U., Silver, J. and Miller, R.H. (1990) Dev. Biol. **138**, 377–390
13. Rudge, J.S. and Silver, J. (1990) J. Neurosci. **10**, 3594–3603
14. Liuzzi, F.J. and Lasek, R.J. (1987) Science **237**, 642–645
15. Johnson-Green, P.C., Dow, K.E. and Riopelle, R.J. (1991) Glia **4**, 314–321
16. Snow, D.M., Watanabe, M., Letourneau, P.C. and Silver, J. (1991) Development **113**, 1473–1486
17. Brittis, P.A., Canning, D.R. and Silver, J. (1992) Science **255**, 733–736
18. McKeon, R.J., Schreiber, R.C., Rudge, J.S. and Silver, J. (1991) J. Neurosci. **11**, 3398–3411
19. Levine, J.M. (1994) J. Neurosci. **14**, 4716–4730
20. Friedlander, D.R., Milev, P., Karthikeyan, L. et al. (1994) J. Cell Biol. **125**, 669–680
21. Bicknese, A.R., Sheppard, A.M., O'Leary, D.D.M. and Pearlman, A.L. (1994) J. Neurosci. **14**, 3500–3510
22. Faissner, A., Clement, A., Lochter, A. et al. (1994) J. Cell Biol. **126**, 783–799
23. Snow, D.M., Steindler, D.A. and Silver, J. (1990) Dev. Biol. **138**, 359–376
24. McCabe, C.F. and Cole, G.J. (1992) Dev. Brain Res. **70**, 9–24
25. Geisert, E.E., Jr., Williams, R.C. and Bidanset, D.J. (1992) Brain Res. **571**, 165–168
26. Bovolenta, P., Wandosell, F. and Nieto-Sampedro, M. (1993) Eur. J. Neurosci. **5**, 454–465
27. Shewan, D., Berry, M., Bedi, K. and Cohen, J. (1993) Eur. J. Neurosci. **5**, 809–817
28. Meiners, S., Marone, M., Rittenhouse, J.L. and Geller, H.M. (1993) Dev. Biol. **160**, 480–493
29. Perez, R.G. and Halfter, W. (1993) Dev. Biol. **156**, 278–292
30. Wehrle, B. and Chiquet, M. (1990) Development **110**, 401–415

31. Bartsch, U., Bartsch, S., Dorries, U. and Schachner, M. (1992) Eur. J. Neurosci. **4**, 338–352
32. Martini, R., Schachner, M. and Faissner, A. (1990) J. Neurocytol. **19**, 601–616
33. Wintergerst, E.S., Fuss, B. and Bartsch, U. (1993) Eur. J. Neurosci. **5**, 299–310
34. Lochter, A., Taylor, J., Fuss, B. and Shachner, M. (1994) Eur. J. Neurosci. **6**, 597–606
35. Taylor, J., Pesheva, P. and Schachner, M. (1993) J. Neurosci. Res. **35**, 347–362
36. Porter, B.E., Weis, J. and Sanes, J.R. (1995) Neuron **14**, 549–559
37. Bray, D., Wood, P. and Bunge, R.P. (1980) Exp. Cell Res. **130**, 241–250
38. Kapfhammer, J.P. and Raper, J.A. (1987) J. Neurosci. **7**, 1595–1600
39. Luo, Y., Raible, D. and Raper, J.A. (1993) Cell **75**, 217–227
40. Kolodkin, A.L., Matthes, D.J., O'Connor, T.P. et al. (1992) Neuron **9**, 831–845
41. Kolodkin, A.L., Matthes, D.J. and Goodman, C.S. (1993) Cell **75**, 1389–1399
42. Püschel, A.W., Adams, R.H. and Betz, H. (1995) Neuron **14**, 941–948
43. Messersmith, E.K., Leonardo, E.D., Shatz, C.J. et al. (1995) Neuron **14**, 949–959
44. Sperry, R.W. (1944) J. Neurophysiol. **7**, 57–69
45. Sretavan, D.W., Feng, L., Puré, E. and Reichardt, L.F. (1994) Neuron **12**, 957–975
46. Cornel, E. and Holt, C. (1992) Neuron **9**, 1001–1011
47. Wizenmann, A., Thanos, S., von Boxberg, Y. and Bonhoeffer, F. (1993) Development **117**, 725–735
48. Drescher, U., Kremoser, C., Handwerker, C. et al. (1995) Cell **82**, 359–370
49. Tessier-Lavigne, M. (1995) Cell **82**, 345–348
50. Stahl, B., Müller, B., von Boxberg, Y., Cox, E.C. and Bonhoeffer, F. (1990) Neuron **5**, 735–743
51. Cox, E.C., Muller, B. and Bonhoeffer, F. (1990) Neuron **4**, 31–37
52. Walter, J., Henke-Fahle, S. and Bonhoeffer, F. (1987) Development **101**, 909–913
53. Wizenmann, A., Thies, E., Klosterman, S., Bonhoeffer, F. and Bähr, M. (1993) Neuron **11**, 975–983
54. Roskies, A.L. and O'Leary, D.D.M. (1994) Science **265**, 799–803
55. Keynes, R.J. and Stern, C.D. (1984) Nature (London) **310**, 786–789
56. Davies, J.A., Cook, G.M.W., Stern, C.D. and Keynes, R.J. (1990) Neuron **2**, 11–20
57. Nose, A., Takeichi, M. and Goodman, C.S. (1994) Neuron **13**, 525–539
58. Tiveron, M.-C., Barboni, E., Rivero, F.B.P. et al. (1992) Nature (London) **355**, 745–748
59. Mahanthappa, N.K. and Patterson, P.H. (1992) Dev. Biol. **150**, 47–59
60. Mahanthappa, N.K. and Patterson, P.H. (1992) Dev. Biol. **150**, 60–71
61. Caroni, P. and Schwab, M.E. (1988) J. Cell Biol. **106**, 1281–1288
62. Schwab, M.E. (1993) Curr. Opin. Neurol. Neurosurg. **6**, 549–553
63. Mukhopadhyay, G., Doherty, P., Walsh, F.S., Crocker, P.R. and Filbin, M.T. (1994) Neuron **13**, 757–767
64. McKerracher, L., Essagian, C. and Aguayo, A.J. (1994) J. Neurosci. **13**, 2617–2626
65. Sagot, Y., Swerts, J.-P. and Cochard, P. (1991) Brain Res. **543**, 25–35
66. Fitzgerald, M., Kwiat, G.C., Middleton, J. and Pini, A. (1993) Development **117**, 1377–1384
67. Pini, A. (1993) Science **261**, 95–98
68. Serafini, T., Kennedy, T.E., Galko, M.J. et al. (1994) Cell **78**, 409–424
69. Kennedy, T.E., Serafini, T., de la Torre, J.R. and Tessier-Lavigne, M. (1994) Cell **78**, 425–435
70. O'Leary, D.D. (1994) Nature (London) **371**, 15–16
71. Hedgecock, E.M., Culotti, J.G. and Hall, D.H. (1990) Neuron **4**, 61–85
72. Hess, D.T., Patterson, S.I., Smith, D.S. and Skene, J.H.P. (1993) Nature (London) **366**, 562–565
73. Mattson, M.P., Dou, P. and Kater, S.B. (1988) J. Neurosci. **8**, 2087–2100
74. Haydon, P.G., McCobb, D.P. and Kater, S.B. (1984) Science **226**, 561–564
75. Lois, C. and Alvarez-Buylla, A. (1994) Science **264**, 1145–1148
76. Carbonetto, S., Evans, D. and Cochard, P. (1987) J. Neurosci. **7**, 610–620
77. Covault, J., Cunningham, J.M. and Sanes, J.R. (1987) J. Cell Biol. **105**, 2479–2488
78. Sandrock, A.W. and Matthew, W.D. (1987) Proc. Natl. Acad. Sci. U.S.A. **84**, 6934–6938
79. Shewan, D., Bedi, K., Berry, M., Winter, J. and Cohen, J. (1994) Neuroprotocols **4**, 142–145
80. Shewan, D., Berry, M. and Cohen, J. (1995) J. Neurosci. **15**, 2057–2062
81. Fitzgerald, M. and Fulton, B.P. (1992) in Sensory Neurons: Diversity, Development and Plasticity (Scott, S., ed.), pp. 287–306, Oxford University Press, New York
82. Colamarino, S.A. and Tessier-Lavigne, M. (1995) Cell **81**, 621–629
83. Cremer, H., Lange, R., Christoph, A. et al. (1994) Nature (London) **367**, 455–459
84. Saga, Y., Yagi, T., Ikawa, Y., Sakakura, T. and Aizawa, S. (1992) Genes Dev. **6**, 1821–1831

9

Electric embryos: the embryonic epithelium as a generator of developmental information

Kenneth R. Robinson and Mark A. Messerli

Department of Biological Sciences, Purdue University,
West Lafayette, IN 47907, U.S.A.

Introduction

Our interest in the possibility that electric fields exist in developing embryos arose from the observation that many cells in culture, especially embryonic cells, respond to surprisingly small applied electrical fields. While there is a substantial and valuable older literature about field effects on cells (reviewed by McCaig [1]), the modern revival of interest dates from the work of Jaffe and Poo [2] on the responses of chick dorsal root ganglion (DRG) neurites. They found that substantial asymmetries in the halo of growing neurites resulted from the application of fields of 70 mV/mm or more and they concluded that the neurites grew faster toward the cathode than toward the anode. The nature of the intact ganglion preparation prevented observations on single neurites; nevertheless, the basic phenomenon was established. That paper was closely followed by two reports that showed that neurites growing from isolated *Xenopus* embryonic neurons responded to applied fields by turning toward the cathode [3,4]. Those papers established that *Xenopus* neurites could sense and respond to a field as small as 7 mV/mm, which corresponds to a voltage drop across their growth cones of less than 500 μV. In addition, both groups found that more neurons differentiated in the presence of a field. Hinkle *et al.* [3] also noted that developing muscle cells showed a striking tendency to form their long axes at right angles to the direction of an applied field. Control experiments made it clear that the electrical field effects were not mediated by induced chemical gradients in the medium, electrode products or temperature changes; rather, the effects were a direct response of the cells to the applied fields.

Since then, the responses of many types of cells to electric fields have been studied. Among these are avian and amphibian neural crest cells [5,6], mammalian neurons [7], embryonic fibroblasts [8], osteoclasts and osteoblasts [9], pollen tubes [10] and fungal zoospores [11]. In a number of cases, the smallest transcellular voltage drop that gave a detectable response was determined and typically was found to be 0.1–1.0 mV per cell diameter, corresponding to a field of 10 mV/mm or less. It is not clear how cells detect and respond to such small fields. One possibility is that transmembrane molecules are asymmetrically redistributed in the plasma membrane, either by direct electrophoresis or by electrically induced water flow near the membrane (electro-osmosis). Both phenomena have been shown to occur (reviewed in [12]), but it has not been possible to connect the redistribution of any particular molecule to a directional cellular response. Furthermore, the fields required to redistribute membrane proteins significantly are larger (by about an order of magnitude) than those that will produce asymmetries of growth or migration. An alternative possibility is that voltage-gated ion channels, particularly calcium channels, might be opened on the depolarized (cathode-facing) side of the cells. Calcium entry on the cathodal side of growth cones has been demonstrated [13,14], but again at much larger fields than the ones that elicit cellular responses.

When this subject was reviewed a decade ago by one of us [15], much of the available information concerned cells from amphibian and avian embryos and it was concluded, "unfortunately, in neither of these cases has it been possible to measure directly either the magnitude or direction of electrical gradients in the interior, and the pathways of the currents through the embryos aren't known". This has been an area of considerable recent progress. We now have information about the size and distribution of electrical fields within both chick and amphibian embryos and we know that disrupting these endogenous fields causes abnormal development, particularly of neural structures. In addition, we know that the neural tube retains or reforms the electrical polarity of the epithelium from which it is formed. The neural tube thus can act as an electrogenic organ and perhaps influence its own development. Here, we will review the evidence for the existence of endogenous electrical fields in embryos and will suggest an additional mechanism by which fields might participate in the emergence of pattern.

Electrical background

In order to follow the ideas about electrical influences on development it is necessary to appreciate a bit of the biophysics of current flow through extended media and the physiology of epithelial transport. While these concepts are not particularly complicated, they are not typically in the forefront of developmental biologists' minds. What follows is a brief primer that is intended to make these matters more accessible to the non-physiologist.

The simplest electrical circuit includes a voltage source (e.g. a battery) and a resistor that is connected to the two terminals of the source by wires that have negligible resistance to current flow. The current carriers in the wires are electrons, which may be imagined to be free, that is, not bound to atoms. For many resistive materials, there is a direct relationship between the voltage difference across them and the amount of current (number of charges per second) that flow through them. The constant of proportionality is the resistance of the resistor. This phenomenological relationship is expressed as Ohm's law: $V = I \cdot R$, where V is the voltage of the battery, I is the current expressed in amperes and R is the resistance in ohms, or alternatively, $I = G \cdot V$, where G, the conductance, is $1/R$.

In an extended medium such as a solution or the interstitial spaces of an embryo, the analysis is analogous but slightly more complex. A major difference is the nature of the charge carriers. In aqueous media, there are essentially no free electrons and electrical charge is carried by ions such as sodium and chloride. While the resistance between any two points may be determined, it is more useful to speak of the bulk resistivity of the medium. Resistivity is expressed in ohm·cm and the resistivity of a typical physiological saline solution is about 100 ohm·cm. If there is a voltage difference between any two points in a conductive medium, there will be a flow of current. The voltage difference per unit distance is the electrical field and is properly expressed in V/m; although we prefer the numerically equivalent unit of mV/mm because it better expresses the scale of voltages and distances of embryos. The relationship between current density and the electrical field is $E = J \cdot \rho$, where E is the electrical field, J is the current density, usually expressed in A/cm^2, and ρ is the resistivity of the medium. The bold type indicates that the electric field and current density are vectors.

In a conductive medium, the existence of an electrical field and the flow of current are inseparable; one implies the other. It is important to realize that the electrical field is a vector; that is, it has both magnitude and direction. It is the directional property of an electric field that makes it a candidate spatial organizer for embryonic development. A question that will be addressed below is whether the magnitude of endogenous fields is sufficient to be an effective organizing agent. Another consideration is the inhomogeneity of embryonic tissue. The endogenous currents flow around and between cells and it is obvious that the local resistivity may vary considerably from place to place, depending on the proximity of cells to each other and the nature of the connections between the cells. The consequence of this is that a given current density may produce quite different electrical fields in different regions, and the scale of these variations may be below the resolution of our present measuring techniques.

In order for currents to flow, there must be a voltage source that has some current-producing capacity. It has long been known that frog skin has considerable ability to generate voltages and drive currents. Lund [16] summarized the results of early work and points out that the cellular layers of the frog skin are arranged so that they produce an inwardly positive potential. An isolated 1 cm^2 of skin can drive currents of tens of microamperes through an external circuit for many hours. The later work of Ussing and colleagues showed that the electrical properties of the frog skin result from the polarized distribution of ion transport machinery [17]. The passive sodium permeability is localized to the outer apical surface while the Na^+/K^+-ATPase is localized on the basal lateral surface. The epithelial cells are joined laterally by tight junctions that limit current flow between the cells and allow for the development of a substantial transepithelial potential (TEP). We now know that the sodium permeability is due to amiloride-sensitive channels that are not voltage-gated. These channels are related to proteins that are involved in neurodegeneration [18]. Frog skin is not the only integument that generates a TEP. Mammals, including humans, maintain significant potentials across their outer epithelial layers with the same polarity as amphibians [19].

If a conductive pathway is provided, the epithelia described above will drive currents and create electrical fields in the adjacent tissues. The most obvious situation where this occurs is a wound to the integument. Fields near epithelial wounds have been measured in several organisms, including cavy skin (140 mV/mm, [20]), amputated newt limbs (60 mV/mm, [21]) and bovine cornea (40 mV/mm, [22]). There is reason to think that these epithelially generated wound currents may be involved in the natural healing process (reviewed by Vanable [23]).

The measurement of currents in the media surrounding cells, embryos and tissues has been made possible by the development of the 'vibrating probe'. Jaffe and Nuccitelli [24] originally invented the method in order to detect the tiny voltage gradients in the sea water near polarizing algal zygotes. The instrument consists of a platinum electrode that oscillates between two points in a conductive medium and if there is current flowing, it will record a voltage difference between the extremes of its vibrational excursion. The signal is then analysed and recorded, and by repositioning the electrode, the current pattern around an organism can be mapped.

Endogenous electrical fields in embryos

Amphibians

As discussed above, the capacity of adult frog skin to generate a TEP and to drive substantial currents has been known for more than a century; however, this capacity was not recognized in embryonic or larval development until considerably later. Early attempts to detect a TEP during embryonic stages failed [25,26] and this failure dampened enthusiasm for the idea that there might be significant endogenous electrical fields in embryos. McCaig and Robinson [27] found that *Xenopus* embryos developed an inwardly positive TEP during the early stages of neurulation and it increased to +60 mV or more during the following hours. This conclusion was challenged by Gillespie [28], who claimed that the detection of substantial TEPs in *Xenopus* embryos was a tip-potential artifact and he put an upper limit of less than 5 mV on any true TEP. Subsequently, Rajnicek *et al.* [29] presented compelling evidence that the TEPs measured by McCaig and Robinson [27] were real; furthermore, they showed that the TEPs were abolished rapidly by

10 μM amiloride or slight damage to the epithelium, and that the embryonic epithelium could drive currents of 20 or 30 μA/cm² through wounds.

Fig. 1. **Vibrating probe measurements around a developing *Xenopus* embryo**

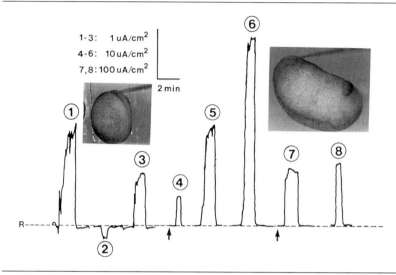

1-3: 1 uA/cm²
4-6: 10 uA/cm²
7,8: 100 uA/cm²

2 min

Upward deflections from the reference line (R) indicate outward current while downward deflections indicate inward current. Arrows between measurements 3 and 4, and 6 and 7 indicate scale changes. Measurement 1 was made between the anterior neural folds at stage 17 and measurement 2 was made lateral to the neural folds near the rostral–caudal midline at stage 17. Measurements 3–8 were made at the blastopore at stages 17, 17½, 18, 19, 20 and 22 respectively. The inset photographs show the vibrating probe in the measuring position at the anterior folds of a stage-17 embryo (left) and at the blastopore of a stage-22 embryo (right). Reprinted from [31] with permission.

The existence of the machinery for producing currents does not establish that currents actually flow through the developing embryo on a global scale. The electrogenic uptake of sodium could be balanced by chloride uptake locally so that the current loops would be small, on the order of cellular dimensions. Using the vibrating probe to map the external current pattern around *Xenopus* embryos, Robinson and Stump [30] found a small inward current over most of the embryonic surface and a locus of considerably larger outward current at the blastopore. It was established that the inward and outward currents were part of the same circuit and that the currents were sodium-dependent and amiloride blockable. Subsequently, Hotary and Robinson [31] used an improved optical arrangement that allowed better positioning of the electrode and found that the blastopore currents were larger than originally reported. The current is detectable at stage 15 and by stage 20, reaches about 100 μA/cm². An example of a vibrating probe record from a single

Xenopus embryo is shown in Fig. 1 and a peak current of 120 $\mu A/cm^2$ was detected at stage 22. Superimposed on these relatively steady currents were brief episodes in which the blastopore currents exceeded 200 $\mu A/cm^2$ (not shown in the Figure). These episodic large currents typically lasted for a minute or two and then returned to the stable level. The steady currents persisted for hours as shown in Fig. 2; the average values shown there do not include episodic currents. In general, it appears that the integument of amphibian embryos is electrically tight and acts as a distributed non-ideal current source while the leaky portions of the integument are highly localized and are the sites of large outward currents.

Fig. 2.   **Average outward current densities measured at the blasto-pore of stage 14–23 *Xenopus* embryos**

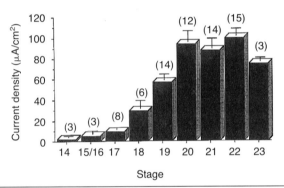

The numbers in parentheses above each bar indicate the number of embryos included in the means. Error bars show the standard error of the mean. Reprinted from [31] with permission.

Given outward currents of the measured magnitude, it seemed likely that there would be measurable voltage gradients within the embryos, but the vibrating probe gives no information about the internal distribution of the currents or the resultant electrical fields. Therefore, conventional glass microelectrodes were used to map the potential at various regions within the embryos. A voltage gradient of 30 mV/mm was measured between the blastopore and points rostral in stage 21–23 embryos, and a gradient of 40 mV/mm was measured between the blastopore and points dorsal [31]. The direction of the field was as expected from the external current measurements: the blastopore was negative with respect to the surrounding tissue.

Another locus of outward current from the *Xenopus* embryo has been detected. At stages 14–17, which is prior to the complete closure of the neural tube, currents of 2–5 $\mu A/cm^2$ left the anterior neural folds [31,32]. Voltage gradients of 10 mV/mm were measured in the vicinity of the anterior neural folds, and again, the direction was consistent with externally detected currents.

Metcalf *et al.* [33] measured striking rostral-caudal voltage gradients beneath the neural plate ectoderm of stage 16 axolotl embryos. The fields averaged about 15 mV/mm (caudally negative) and were sometimes as large as 40 mV/mm. The externally measured currents were transiently dependent on sodium and blockable by amiloride, but the currents recovered within about 2 h even in the absence of sodium or presence of amiloride. Metcalf *et al.* [33] termed these 'adaption currents' and their ionic basis is not known. A similar situation has been noted with regard to the TEP and blastopore currents in *Xenopus* embryos [31]; that is, the effects of channel blockers or sodium removal is transient.

Avian embryos

A second embryonic system in which the existence of developmental currents and significant voltage gradients has been measured is the domestic chicken. Jaffe and Stern [34] inferred the existence of 100 $\mu A/cm^2$ leaving the primitive streak of embryos. We say 'inferred' because they were unable to position the vibrating electrode near the primitive streak and thus had to estimate the magnitude of the surface currents from measurements made relatively far away. Stern and MacKenzie [35] attempted to measure the resultant voltage gradients within the embryos, but were unsuccessful. Hotary and Robinson [36] studied older embryos and found that the posterior intestinal portal of the developing hind gut was an intense locus of outward current. These currents were detectable at stage 15, continued through stage 22, and were maximal at stage 17 when they averaged more than 110 $\mu A/cm^2$ at the site of the electrode. Unlike the situation with the primitive streak currents [34], the large currents at the posterior intestinal portal were measured directly, not inferred from measurements done some distance away. Other regions of the embryos were mapped at those stages, and only small inward currents were found.

In view of the magnitude of the currents, we searched for intraembryonic voltage gradients using microelectrodes. Significant electrical fields averaging more than 20 mV/mm (caudally negative) were found in the caudal third of the embryos, and in some cases, were more than 30 mV/mm. These were the first measurements of electrical fields in embryos; the amphibian results discussed above were later findings.

The underlying basis for the posterior intestinal portal currents and the resultant electrical fields within the chick embryo was described more than 70 years ago. As we have already discussed, the creation of currents on the scale of the whole embryo requires a voltage source (the integument's TEP) and a localized low-resistance pathway. Boyden, in 1922 [37], first described the breakdown of the intestinal epithelium that occurs during tail gut reduction and this observation was later confirmed by Zwilling [38]. During this process, the epithelial lining of the cloacal wall disintegrates, leaving the underlying mesenchyme exposed to the subgerminal fluid. The currents that we measure are the physiological manifestation of this developmentally programmed shunt through the integument. The timing of the intestinal degeneration given by Boyden [37] closely matches the timing of the posterior intestinal portal currents [36].

Summary

It is now evident that the embryos of many, perhaps all, vertebrates develop an electrically polarized outer epithelium quite early in development. The polarity of the TEP is always the same: internally positive. In the case of *Xenopus*, there is already a high-resistance seal between the blastomeres by the early blastula stages and a TEP of 20 mV, inside positive, is supported [39]. In addition, a specific pattern of developmentally regulated leaks in the integument occurs, producing currents that flow inwardly through much of the embryo, out the leak, and return through the surrounding medium. In the chicken embryo, these outward currents are first associated with the primitive streak and later with the developing posterior gut [34,36]. In *Xenopus*, the locus of outward current is first at the anterior neural plate, then the blastopore [31] and later at the site of limb bud formation [40]. Many of the features of the *Xenopus* pattern are shown by axolotl embryos [33,41]. Mammalian embryos have been less well studied from this perspective, but the 7-day mouse embryo produces up to 60 $\mu A/cm^2$ of outward current along the midline of the embryo [42].

What is the developmental significance of the widespread occurrence of active, electrogenic transport of ions into embryos with the resultant formation of a TEP? In amphibians, which develop in a dilute and uncertain environment, it might be assumed that the primary role is to accumulate sodium as well as initially to drive the expansion of the blastocoel by osmotic means. While this may be true with regard to blastocoel formation, it is not true that *Xenopus* or axolotl embryos require sodium uptake. If deprived of external sodium or bathed in amiloride, most embryos develop normally. Furthermore, the TEP and the associated currents persist and increase long after blastocoel formation is complete. Embryos grown in the absence of external sodium or in the presence of amiloride also re-establish a TEP of the usual polarity and restore their endogenous currents to nearly normal levels, perhaps by active inward transport of calcium. In chick, there is even less reason to believe that the inward transport of positive ions has to do with salt accumulation as the bathing fluid is not very different from the interstitial fluid.

It is a reasonable hypothesis that the function of the robust and widely conserved practice in vertebrate embryos of developing an outer epithelium that can both establish an inwardly positive TEP and drive substantial currents through developmentally programmed leaks is to produce intraembryonic electrical fields. Recent measurements reveal that electrical fields of tens of millivolts per millimeter exist in embryos and are produced by the internal loops of the externally detected currents. The pattern of these fields is stage- and species-dependent, but invariant for a given embryonic stage of a particular species. The direction of the currents is aligned with the major axes of the embryos. Possible targets of the fields will be discussed below.

Effects of disrupting endogenous fields

There are now three well-documented cases in which the normal pattern of embryonic current has been disrupted and the effects on development noted. The first experiments of this kind were performed on chick embryos by Hotary and Robinson [32], who used a passive approach in the form of an electrical shunt. (An active approach would be to pass current through the embryo from an external source.) The strategy was to create a small ectopic leak in the integument in order to reduce the currents along the normal pathways and to make a new path for current. As embryonic skin rapidly heals small wounds, it was necessary to devise a method for keeping the wound open. The solution was to implant fine, saline-filled pipettes into slits in the flanks of stage 11–14 embryos. The pipettes were about 1 mm long and had a diameter of 75–100 μm. In some cases the filling solution was gelled with agarose to prevent bulk flow of fluids. Access to the embryo was gained through a window in the shell, which was then closed to allow subsequent development. Control embryos were treated in the same way, except that they received a solid glass implant. Vibrating probe measurements verified that the shunts were sources of outward currents (Fig. 3) and the solid glass implants were not.

Fig. 3. **Two-dimensional vibrating probe measurements of currents from an implanted shunt in a chick embryo**

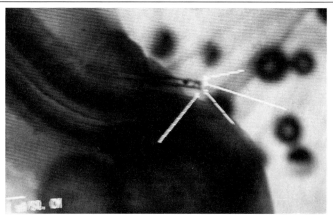

Current vectors are shown as lines leaving a dot that indicates the electrode's position at the time of the measurement. The line's length is proportional to the current density and the scale bar in the left-hand corner represents 1 μA/cm². These measurements were made about 30 h after implanting the shunt. Reprinted from [32] with permission.

The effects of implanted shunts were striking. More than 90% of the embryos that received shunts developed externally apparent abnormalities, while only 11% of the controls were abnormal. The abnormalities involved the tail, the limbs, and the head, the most common being at the tail. Examples of some of these abnormalities are shown in Fig. 4. Often, tail development stopped altogether; in other cases, the notochord continued to extend but turned 180° rostrally and grew into the cloaca (Fig. 4D). Sometimes, a nipple-like projection formed that was nearly as long as the normal tail, but these structures lacked the usual organized elements of somites, neural tube and notochord, and instead were sacs of disorganized mesenchyme. Scanning electron micrographs of embryos with solid and open implants are shown in Fig. 5.

Fig. 4.   **Abnormalities in limb, head and gut development in current-shunted chick embryos**

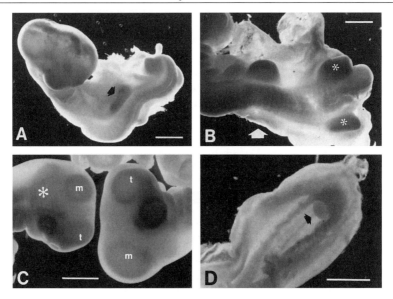

(A) Bilobed wing bud (arrow) on the side contralateral to the implant. Note also the abnormal tail. (B) The ipsilateral wing bud is completely absent (arrow) and both leg buds are enlarged and flattened distally (asterisks). (C) Brain development was retarded in the current-shunted embryo on the left (asterisk). Only the mesencephalon (m) and the telencephalon (t) are labelled, but all of the brain divisions were abnormal. The embryo on the right also received an active current shunt and developed tail abnormalities (not shown), but had normal brain development. (D) Ventral side of an experimental embryo showing the abnormal outgrowth of the notochord into the gut and emerging from the posterior intestinal portal (arrow). Scale bars: in A, 1 mm; in B, C and D, 0.5 mm. Reprinted from [32] with permission.

Treated embryos often had abnormal limb development. In some cases, one or more limb buds were completely absent while in other cases, limb buds were flattened or bilobed. The site of the limb abnormalities was not correlated with the location of the shunt in any obvious way, but leg buds were affected more frequently than wing buds. A smaller fraction of the treated embryos (but none of the controls) developed abnormal brain structures. The various divisions of the brain were reduced in size, as was the whole head, in affected embryos.

Fig. 5.   **Scanning electron micrographs of experimental and control chick embryos**

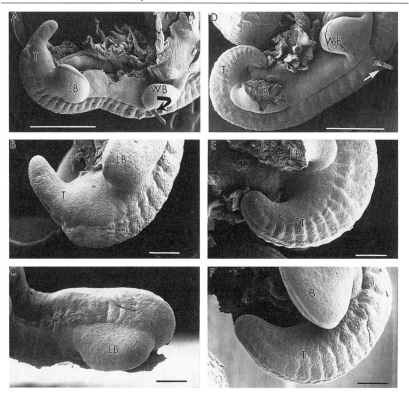

(A) and (D) show low magnification views of current-shunted (A) and solid glass rod-implanted control (D) embryos. The arrows in each indicate the position of the implants. (B) Higher magnification view of the tail region of the embryo shown in (A). The tail (T) is nearly normal in length, but the distal half appears unstructured, lacking any somites or extension of the neural tube. (C) Experimental embryo that entirely failed to develop a tail, terminating in a blunt stump just distal to the leg buds (LB). (E) Tail region of the solid glass rod control shown in (D). (F) Unoperated control showing normal tail. Scale bars: in A and D, 1 mm; in B, C, E and F, 0.25 mm. Reprinted from [32] with permission.

The effects of the shunts on embryonic development were so dramatic and widespread that our preconceived notions about the mechanism were challenged. We had anticipated fairly subtle effects on caudal development, perhaps caused by misdirected neural crest cells, and we had thought that Remak's ganglion might be a likely target. Instead, the developmental abnormalities extended along the entire embryonic axis, although they showed a caudal–rostral gradient in frequency. The tail was affected most often, followed by the leg bud, the wing bud, and the head, in order of frequency. This caused us to suggest that the endogenous fields might be affecting the distribution of diffusible morphogens [32], an idea that we will pursue below. It is important to emphasize that the shunts and the solid glass control implant did not cause any local tissue breakdown or other abnormalities at the site of implantation.

The embryos that were defective in tail development resembled a well-known mutant of chickens, *rumpless*. This mutation was characterized extensively by Zwilling [38,43]. We were curious to know if mutant embryos were physiologically defective in their ability to generate a TEP or to drive currents. We found that *rumpless* embryos exhibiting the mutant phenotype had hind gut currents that were about one-half as large as those of wild-type embryos [32]. One possible explanation of these results is that the tail abnormalities in our shunted embryos and in the *rumpless* embryos were due to decreased currents leaving the posterior intestinal portal, while the limb and head abnormalities in the shunted embryos were due to the presence of an ectopic leak, a situation that does not occur in the mutants.

More recently, there have been two reports of the effects of actively modifying the endogenous currents patterns of amphibian embryos. Metcalf and Borgens [44] grew axolotl embryos in the presence of external electrical fields that were created by passing current through artificial pond water in which the embryos were bathed. With fields in the range of 25–75 mV/mm applied to neurula-stage embryos, they saw an astonishing array of defects that depended on which end of the embryo faced the cathode. If the rostral ends of the embryos faced the cathode of the external circuit, head defects predominated and if the caudal ends faced the cathode, lower abdominal and tail defects predominated. The head defects included the formation of mesenchyme-filled blisters, duplicated lumens of brain structures, the presence of cells within lumens of the brain and spinal cords, and the absence of recognizable brain structures.

Hotary and Robinson [31] employed a different strategy for actively modifying endogenous fields in *Xenopus* embryos. They impaled the embryos with glass microelectrodes that were similar to the ones used to measure the TEP and used them to pass current from an external reference electrode so that the endogenous current leaving the blastopore was reduced or reversed. Controls included passing no current through the impaling electrode or modestly augmenting the endogenous blastopore current. The effect of the applied currents on the blastopore currents was monitored directly by the vibrating probe. An applied current of 100–200 nA nulled the endogenous current and 500 nA approximately reversed the endogenous currents of stage-21 embryos, and when this was applied from stage 15 to stage 23 embryos developed highly abnormally. A common early response was that cells streamed out of the blastopore (Fig. 6A). Later effects were the formation of ventral bulges, some of which appeared to develop into ectopic cement glands and greatly reduced or absent brain structures (Fig. 6). There was an interesting additional defect. *Xenopus* embryos become ciliated and will glide forward along the surface due to the beating of the cilia. This is especially apparent in older embryos that have been anaesthetized so that they do not swim. Current-treated embryos failed to glide at all or did so more slowly than normal. The cilia were either absent or non-functional as a consequence of the current treatment.

Fig. 6.   The effects of internally applied electrical currents on *Xenopus* embryonic development

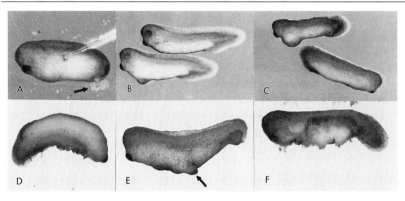

(A) An embryo with the current-passing electrode in place. The current was 100 nA, which approximately nulled the endogenous current at the blastopore. The arrow indicates cells that were extruded from the blastopore, a common response to currents. (B) The treated embryo shown in (A) (below) and a sibling control embryo that was impaled but received no current. The pigmented bulge on the ventral side of the treated embryo was a common feature. (C) Another treated (above) and untreated pair of embryos. The current was 100 nA. (D) and (E) Embryos that received 250 nA of current. Note reduced or absent head structures. The arrow in (E) indicates an ectopic cement gland, as judged by the fact that it became sticky. (F) An embryo that received 500 nA, which approximately reversed the normal outward current at the blastopore. It was disintegrating all along its ventral side. Reprinted from [31] with permission..

It should be pointed out that implanting electrodes to reverse or null the blastopore currents is not equivalent, in the interior of the embryos, to stopping or reversing the endogenous current. The impaling electrode is an intense local sink for current and will have a complicated, uncertain effect on the global field pattern within the embryo. Because of the opacity of the embryos, the location of the tips of the impaling electrodes cannot be known with certainty and may not always lie in the same compartment. Nevertheless, this method surely does allow the endogenous fields within the embryos to be disrupted, even if the details of the disruption are not known.

Despite the variety of organisms and the different methods used, disrupting endogenous embryonic electrical fields seems to produce a surprisingly uniform array of developmental defects. Brain structures seem to be especially vulnerable to these manipulations and the deformities produced are quite similar in the different organisms (c.f. Metcalf and Borgens [44], Fig. 2, and Hotary and Robinson [31], Fig. 6D). Likewise, caudal development is also sensitive to alterations in the normal field pattern. These two sites, i.e. the head and the tail, are also the loci of endogenous outward currents and are the regions where detectable electrical fields occur.

The neural tube as an electrogenic structure

The neural tube of vertebrates is derived directly from the neural plate ectoderm. The lateral edges of the neural plate thicken and elevate, and then move toward the dorsal midline where they fuse. The hollow tube thus formed detaches from the overlying ectoderm and becomes the neural tube. Topologically, the lumen of the neural tube is equivalent to the embryonic exterior. We were curious to know if the neural tube cells retained the polarity of the parent epithelium, and if so, what the physiological consequences of that might be.

It turned out to be reasonably easy to insert microelectrodes into the lumen of the recently closed neural tubes of *Xenopus* embryos and to record electrically from them [45]. However, it must be remembered that an electrode in the lumen of the neural tube reports the voltage difference between itself and an external reference, and that involves two epithelial layers: the ectoderm and the neural tube itself. It was necessary to measure the TEP (i.e. the potential across the ectoderm) near the site of the neural tube impalement and then to subtract that value from the reading of the neural tube electrode. We refer to that difference as the transneural tube potential (TNTP). The neural tube does in fact generate a potential between its lumen and the interstitial fluid, with the lumen negative with respect to the fluid. This TNTP is largest at stage 23 when it averages −23 mV after which it declines to −14 mV by stage 25.

It was important to demonstrate that the recording electrodes were actually in the neural tube. In some cases, we included fluorescein isothiocyanate (FITC)-labelled concanavalin A (Con A) in the recording electrode and pressure-injected a small bolus after a stable recording was achieved. These embryos were later fixed and transversely sectioned, and the sections examined with fluorescence microscopy. Con A staining was seen only on the lumenal surface of the neural tube. The results of those experiments confirmed that the electrical criterion for impalement of the neural tube (a sudden negative change in the TEP as the electrode was advanced) was adequate.

It was also possible to advance the electrode through the ventral surface of the neural tube once a TNTP reading was obtained. This resulted in a sudden positive shift in the potential. Quite to our surprise, the TEP recorded there was more positive than the TEP measured just lateral to the neural tube by an average of 5 mV. In no other location was such a large voltage difference seen over such a short distance in the absence of an epithelial layer. It suggests that the neural tube drives currents through itself in a dorsal–ventral direction with a ventral–dorsal return loop. The electrical field thus created around the neural tube is 50–100 mV/mm. These results also imply that the properties of the ventral part of the neural tube are different from those of the dorsal region, either in cellular coupling, ion transport properties, or both.

Recently, Shi and Borgens [46] have investigated the electrical aspects of the neural tube of the axolotl embryo. They report that the lumen of the neural tube is a startling 90 mV negative with respect to the interstitial space at stage 28. In addition, they iontophoresed amiloride and benzamil into the lumen and found that the TNTP was collapsed. This treatment also led to severe abnormalities of cranial development.

What are the implications of the fact that the neural tube is an active, ion-transporting epithelium that maintains a substantial potential across itself? One might first think that it is a necessary consequence of the need to modify the dilute fluid with which it is filled when it is first formed. This view is difficult to reconcile with the results of Shi and Borgens [46] who found that amiloride applied to the lumenal surface eliminated the TNTP. If the basis of the TNTP is sodium transport out of the lumen, then the system is not serving the function of concentrating the lumenal fluid; presumably that is carried out by some other mechanism. On the other hand, it is inescapable that the lateral surfaces of the cells of the neural tube are exposed to huge electrical fields. As the walls of the *Xenopus* neural tube are about 40 μm thick, a TNTP of 20 mV corresponds to a field of 500 mV/mm. The fields across the axolotl neural tube are even larger. Fields of this magnitude are easily capable of redistributing proteins in the plane of the membrane [12] and such redistributions may be involved in the differentiation of neurons. In this sense, the neural tube may be seen as a self-polarizing structure where the initial polarity of the neural tube leads to the formation of a TNTP, which then affects differentiation.

Another aspect is the fact that the TNTP is not radially uniform. There are two consequences of this. First, the ventral aspect of the neural tube has a larger potential across it than the more dorsal regions, which may contribute to the dorsal–ventral spatial differentiation of the neural tube. Secondly, the resultant current around and through the neural tube may act to affect the distribution of developmentally important molecules within and around the neural tube. This later possibility will be considered further below.

Redistribution of morphogenetically important molecules by endogenous electrical fields

The foregoing discussion has been based on the presupposition that the likely mode of action of endogenous electrical fields on development is by way of direct guidance of cells. In view of the unexpectedly broad effects of disrupting endogenous fields, we have begun to consider an additional possibility. The idea of gradients of diffusible morphogens being involved in embryonic patterning has a long history and recent advances have implicated specific signalling molecules, such as the activins and Sonic hedgehog, that are secreted by one group of cells and diffuse to other groups where they convey developmental information. Crick [47] has presented a valuable analysis of the time required to set up gradients by diffusion, and he concludes that gradients could well be established over biologically relevant distances in reasonable times. However, diffusion is an inherently symmetrical process and the gradient produced around a source will be the same in all directions. If different responses to a uniformly diffusing morphogen are observed in different directions from the source, it must be because there are pre-existing differences in the cells that encounter the morphogen, which begs the question of how those differences are set up in the first place. The possibility that gradients are involved in embryonic pattern formation would be more attractive if diffusion could be directed.

We have begun to study the effects of electrical fields on the diffusion of proteins. Due to the practical difficulty of using actual candidate morphogens, we have used BSA in these preliminary experiments. It is useful for these studies because it is readily available, it is well-characterized in terms of its diffusion coefficient and its electrophoretic mobility, and it is a fairly typical protein in these regards. The question we wish to answer is whether physiological electrical fields (10–100 mV/mm) can affect the pattern of diffusion in a significant way.

The net movement of a molecular species through a plane is called a flux (J) and it has the units of mol/(cm^2·s). If there is a net flux across a particular plane, there must be a gradient in the concentration and the relationship between the flux and the gradient is given by $J = D \cdot dC/dx$, in one dimension, where D is the diffusion coefficient, which has units of cm^2/s, and C is the concentration in mol/cm^3. If the molecular species is charged and if there is an electrical field present, there will be an additional movement that is superimposed on diffusive movement. An electrical field can create or maintain a gradient, while diffusion acts inexorably to limit or destroy gradients. It is instructive to consider a simple, if somewhat unrealistic example. What electrical field is required to maintain a twofold gradient of BSA over a distance of 200 µm? The electrophoretic flux in one direction will equal the diffusion-driven flux in the other direction when $CE\mu = D \cdot \Delta C/\Delta x$, where C is the concentration, E the electric field and μ the electrophoretic mobility. For BSA at physiological pH, D is 6×10^{-7} cm^2/s and $\mu \cong -1.0 \times 10^{-4}$ cm·s^{-1}·V^{-1}·cm^{-1}. Using these values in the equation, solving for E gives a field of 0.20 V/cm or 20 mV/mm.

Fig. 7. **The effect of electric fields on the diffusion of FITC–BSA**

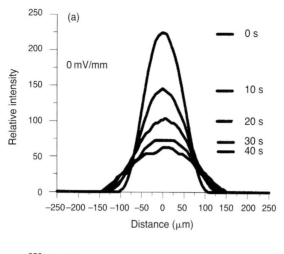

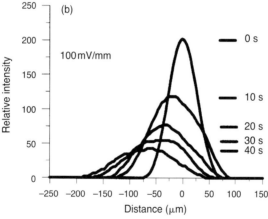

FITC–BSA in 100 mM NaCl was pressure injected into a 0.5% agarose gel at pH 7.3. The 100–150 mm diameter bolus was imaged while it diffused through the gel. A profile was then taken through the centre of each image indicating the fluorescent intensity as a function of position. The bolus was injected from the electrode at 0 μm. **(a)** Shows the profiles for diffusing FITC–albumin without an electric field for a 40 s time period. The peak intensity decreases as the albumin diffuses throughout the gel but remains symmetrical about the injection point, 0 μm. **(b)** Shows the profiles for diffusing FITC–albumin in the presence of a 100 mV/mm electric field. The field was parallel to the x-axis with the anode/cathode polarity corresponding to the negative/positive numbers on the number line. The peak intensities over time, again decrease as diffusion occurs, but in the presence of the electric field migrate toward the anode. The profiles are no longer symmetrical about the injection point but are symmetrical about points in the direction of the anode.

We assume here that the concentration varies linearly over the 200 μm region. Thus, the prediction of this exercise is that modest electrical fields, toward the low end of what are measured in embryos, can maintain a significant gradient of BSA against the levelling action of diffusion over biologically relevant distances. It should be emphasized that BSA is in no way an unusual protein with respect to its diffusion coefficient and electrophoretic mobility. The diffusion coefficient of a spherical molecule varies inversely with the radius and thus inversely with the cube root of the molecular mass. Electrophoretic mobilities vary widely and are positive for basic proteins, but BSA is quite typical.

As a test of these considerations, we have directly imaged the movement in two dimensions of FITC–BSA following the injection of a small bolus into an agarose gel (pH 7.3), using quantitative video-microscopy. In the absence of an electrical field, we observe symmetrical diffusion. When the movement of BSA is monitored in the presence of an electrical field, the electrophoretic drift toward the anode can be seen, superimposed on diffusion. An example of this phenomenon at a field of 100 mV/mm is shown in Fig. 7b. We detect significant deviations from symmetrical diffusion at fields at least as small as 25 mV/mm, in good agreement with the calculations outlined above. We conclude that endogenous electrical fields in embryos may affect development by regulating the distribution of morpho-genetic molecules.

Direct evidence for electric control of protein distribution in a biological system exists. The oocytes of certain insects develop from a series of incomplete cell divisions that results in each oocyte being connected to seven nurse cells by cytoplasmic bridges. In the case of the moth, *Hyalophora cecropia*, the bridges are about 30 μm wide and 50 μm long. Proteins and RNA made in the nurse cells cross the cytoplasmic bridges into the growing oocyte. Woodruff and Telfer [48] have shown that these bridges are electrically polarized, with the nurse cells being about 6 mV negative with respect to the oocyte, corresponding to an electrical field of 100 mV/mm across the bridge. In a remarkable set of experiments, they then explored the consequences of this electrical field on the distribution of proteins [49]. They injected a highly positively charged protein, lysozyme, into nurse cells and oocytes, and they found that the protein diffused freely from the oocyte into the nurse cells but never from the nurse cells into the oocyte to any measurable extent. They then modified lysozyme by methylcarboxylation, rendering it negatively charged at physiological pH. The modified protein crossed the cytoplasmic bridges in the opposite way, i.e. from nurse cell into oocyte, but not from oocyte to nurse cell. These experiments provide strong evidence that electrical fields can regulate the distribution of charged molecules in developing biological systems.

Other considerations

There are other possible electrical fields in developing embryos besides those generated by epithelia. The growing neurite may drive currents through itself and produce longitudinal fields within the axoplasm. An extreme and somewhat artificial example of this situation occurs when an axon is severed. The membrane potential collapses at the open end of the proximal segment but a few length constants toward the cell body, the membrane potential is nearly normal and a substantial voltage gradient — an electric field — is set up in the intervening distance. In a large axon such as the Mauthner axon of the larval lamprey, such fields can be measured directly and they may be as large as 50 mV/mm [50]. The growth cone of the growing axon may also be more permeable to ions, particularly calcium, than the more proximal parts of the axon and thus depolarized. An electrical field would then exist along the axon and could affect the alignment of microtubules and the distribution of organelles such as mitochondria. These hypothesized intracellular electrical fields might be one of the targets of externally applied fields as an applied field parallel to a growing axon will penetrate to the interior, to at least some extent.

In any case, it is now evident that steady electrical fields, persisting for hours or even days, exist at specific places within vertebrate embryos. These fields are considerably larger than those that produce easily detectable responses by embryonic cells *in vitro*. The generators of these fields are mainly the epithelial layers, particularly the ectoderm, that are among the earliest organs to form during embryogenesis. The available evidence indicates that disruption of endogenous fields severely disturbs development. In addition to direct effects on cells, the endogenous fields are large enough to affect the distribution of the secreted molecules that may act as morphogens. The task confronting developmental biologists concerned with electrical fields as organizing agents is to identify specific targets of endogenous fields, which will require subtler means of perturbing the fields, or perhaps selectively restoring fields to show rescue of particular defects.

References
1.   McCaig, C.D. (1988) Progr. Neurobiol. **30**, 449–468
2.   Jaffe, L.F. and Poo, M.-M. (1979) J. Exp. Zool. **209**, 115–128
3.   Hinkle, L., McCaig, C.D. and Robinson, K.R. (1981) J. Physiol. **314**, 121–135
4.   Patel, N.B. and Poo, M.-M. (1982) J. Neurosci. **2**, 483–496
5.   Stump, R.F. and Robinson, K.R. (1983) J. Cell Biol. **97**, 1226–1233
6.   Gruler, H. and Nuccitelli, R. (1991) Cell Motil. Cytoskel. **19**, 121–133
7.   Cork, R.J., McGinnis, M.E., Tsai, J. and Robinson, K.R. (1994) J. Neurobiol. **25**, 1509–1516
8.   Erickson, C.A. and Nuccitelli, R. (1984) J. Cell Biol. **98**, 296–307
9.   Ferrier, J., Ross, S.M., Kanehisa, J. and Aubin, J.E. (1986) J. Cell. Physiol. **129**, 283–288
10. Wang, C., Rathore, K.S. and Robinson, K.R. (1989) Dev. Biol. **136**, 405–410
11. Morris, B.M. and Gow, N.A.R. (1993) Phytopathology **83**, 877–882
12. Poo, M.-M. (1981) Annu. Rev. Biophys. Bioeng. **10**, 245–276
13. Bedlack, R.S., Jr., Wei, M.-D. and Loew, L.M. (1992) Neuron **9**, 393–403
14. Davenport, R.W. and Kater, S.B. (1992) Neuron **9**, 405–416

15. Robinson, K.R. (1985) J. Cell Biol. **101**, 2023–2027
16. Lund, E.J. (1947) Bioelectric Fields and Growth. University of Texas Press, Austin
17. Koefoed-Johnsen, V. and Ussing, H.H. (1958) Acta Physiol. Scand. **42**, 298–308
18. Canessa, C.M., Horisberger, J. and Rossier, B.C. (1993) Nature (London) **361**, 467–470
19. Barker, A.T., Jaffe, L.F. and Vanable, J.W. (1982) Am. J. Physiol. **242**, R358–R366
20. Jaffe, L.F. and Vanable, J.W. (1984) Clin. Dermatol. **2**, 34–44
21. McGinnis, M.E. and Vanable, J.W. (1986) Dev. Biol. **116**, 184–193
22. Chiang, M., Robinson, K.R. and Vanable, J.W. (1992) Exp. Eye Res. **54**, 999–1003
23. Vanable, J.W., Jr. (1989) in Electric Fields in Vertebrate Repair (Borgens, R.B., Robinson, K.R., Vanable, J.W., Jr. and McGinnis, M.E., eds.), pp. 171–224, A. R. Liss, New York
24. Jaffe, L.F. and Nuccitelli, R. (1974) J. Cell Biol. **63**, 614–628
25. Alvarado, R.H. and Moody, A. (1970) Am. J. Physiol. **218**, 1510–1516
26. Taylor, R.E., Jr. and Barker, S.B. (1965) Science **148**, 1612–1613
27. McCaig, C.D. and Robinson, K.R. (1982) Dev. Biol. **90**, 335–339
28. Gillespie, J.I. (1983) J. Physiol. **244**, 359–377
29. Rajnicek, A.M., Stump, R.F. and Robinson, K.R. (1988) Dev. Biol. **128**, 290–299
30. Robinson, K.R. and Stump, R.F. (1984) J. Physiol. **352**, 339–352
31. Hotary, K.B. and Robinson, K.R. (1994) Dev. Biol. **166**, 789–800
32. Hotary, K.B. and Robinson, K.R. (1992) Development **114**, 985–996
33. Metcalf, M.E.M., Shi, R. and Borgens, R.B. (1994) J. Exp. Zool. **268**, 307–322
34. Jaffe, L.F. and Stern, C.D. (1979) Science **206**, 569–571
35. Stern, C.D. and MacKenzie, D.O. (1983) J. Embryol. Exp. Morphol. **77**, 73–98
36. Hotary, K.B. and Robinson, K.R. (1990) Dev. Biol. **140**, 149–160
37. Boyden, E.A. (1922) Am. J. Anat. **30**, 163–201
38. Zwilling, E. (1942) Genetics **27**, 641–656
39. Regen, C.M. and Steinhardt, R.A. (1986) Dev. Biol. **113**, 147–154
40. Robinson, K.R. (1983) Dev. Biol. **97**, 203–211
41. Borgens, R.B., Rouleau, M.F. and DeLanney, L.E. (1983) J. Exp. Zool. **228**, 491–503
42. Winkel, G.K. and Nuccitelli, R. (1989) Biol. Bull. Mar. Biol. Lab., Woods Hole **176**(S), 110–117
43. Zwilling, E. (1945) J. Exp. Zool. **99**, 79–91
44. Metcalf, M.E.M and Borgens, R.B. (1994) J. Exp. Zool. **268**, 323–338
45. Hotary, K.B. and Robinson, K.R. (1991) Dev. Brain Res. **59**, 65–73
46. Shi, R. and Borgens, R.B. (1994) Dev. Biol. **165**, 105–116
47. Crick, F. (1970) Nature (London) **225**, 420–422
48. Woodruff, R.I. and Telfer, W.H. (1973) J. Cell Biol. **58**, 172–188
49. Woodruff, R.I. and Telfer, W.H. (1980) Nature (London) **286**, 84–86
50. Strautman, A.F., Cork, R.J. and Robinson, K.R. (1990) J. Neurosci. **10**, 3564–3575

10

Nerve growth and nerve guidance in a physiological electric field

Colin D. McCaig† and Lynda Erskine*

Department of Biomedical Sciences, Marischal College, University of
Aberdeen, Aberdeen AB9 1AS, Scotland, U.K.

Introduction

On christening the growth cone and describing it as "a sort of club or battering ram, endowed with exquisite chemical sensitivity .... and with a certain impulsive force..", Cajal, characteristically, was ahead of his time in realizing that growth cones would respond to several environmental influences. Papers and textbooks of the time indicate that extracellular direct current (DC) electric fields were regarded as respectable, likely players in shaping neuronal morphology [1,2]. Indeed, among the earliest tissue-culture experiments (from Harrison's laboratory, where the technique originated), was a demonstration that a small DC electric field caused neurites to grow "almost entirely along the lines of force in the galvanic field" [3]. While other contemporaneous hypotheses now form the core of widely held views regarding nerve guidance mechanisms, many fewer students of the growth cone either know that DC fields can guide nerve growth, or accept that this may be physiologically relevant. This overview will attempt to redress this unawareness and scepticism. The strategy is to show not only that DC fields alone have an impressive array of effects on nerve growth and guidance, but to use as circumstantial evidence the increasing indications that DC fields also interact in important ways with co-existent 'respectable' cues, to modulate nerve growth and guidance. Additionally, we shall show that the underlying mechanisms involve elements of central importance in neuronal functioning, making it unlikely that electric field-induced nerve guidance is an epi-phenomenon. Parallel supporting arguments and evidence indicating that DC fields exist in developing and regenerating systems, and that disrupting these fields disrupts neuronal development are reviewed in Chapter 9 and will be covered only briefly here.

Present address: Department of Anatomy and Developmental Biology, University College London, Windeyer Building, Cleveland Street, London W1P 6DB, U.K.
†*To whom correspondence should be addressed.*

Electric field-induced nerve growth and guidance: a historical perspective

As indicated above, Ingvar's work [3] was the earliest experimental demonstration that an applied electric field could influence nerve growth. This work had limited impact, only ever appearing as a short abstract, without diagrams or quantification of the evidence. These were not the days of striving for high impact journals! There was further tantalization in this report; cell processes growing towards the anode were morphologically different from those which projected towards the cathode. Ingvar may have been suggesting that axons and dendrites differ in their responsiveness to an applied electric field. Seventy three years later we find that he was correct [4] (see below).

In the intervening 60 years there were only two relevant publications of consequence [5,6]. Weiss was unequivocal in denying that electrical fields (and incidentally chemotropic influences) affected neuronal growth. He is worth quoting. "For one and a half years ceaseless attempts were made to influence nerve fibres growing *in vitro* by electric currents: however not the remotest indication of any success was ever detected. ...it can be regarded as almost certain that electric currents are not among the agents orienting the course of nerve fibres". By contrast, Marsh and Beams [6] provided powerful evidence, carefully documenting that outgrowths from chick medullary explants (7–10 day embryos) grew selectively towards the cathode and were suppressed on the anodal side of the explants. Threshold levels were 64 mV/mm and 53 mV/mm for cathodal orientation and anodal suppression respectively. A crucial additional finding was that fibres emerging from the explant at an angle to the current axis turned "in more or less continuous curves to grow towards the cathode". Nonetheless, adequate controls which would have established beyond doubt that these effects were attributable to the electric field itself and not to a secondary influence of the field had not been performed. An electric field might establish a chemical gradient by electrophoresis of charged macromolecules in the culture medium; neurites or cell bodies might be physically moved by electrophoresis or by electro-osmotic movements of culture fluid; or molecular elements within the substrate might be reorganized by the electric field to form preferred physical pathways, as Weiss had suggested [5]. These issues have been tackled in more recent studies.

Electric field-induced nerve growth and guidance: the modern period

In the late 1970s it emerged that not only might the direction of neurite growth be determined by an electric field, but that neurite growth rate also could be influenced. (The growth rate of neurites *in vivo* has an underemphasized influence in patterning a nervous system. Axons of chick cranial sensory neurons originating in discrete ganglia must grow differing distances to reach their appropriate target. Axons with more distant targets grow faster, although here this is an intrinsic property of the axons, since it is seen also in cultured neurons [7]). Sensory neurites from dorsal root ganglion (DRG) explants (7–9 day chick embryo) grew around three times faster towards the cathode than the anode in DC fields between 70 and 140 mV/mm. [8]. The DRG cell mass showed no net movements, while convincing theoretical arguments were brought forward to refute the possibility of induced chemical gradients or molecular reorganization of the substrate accounting for this observation. Significantly, no turning of neurites to grow cathodally was seen.

This work provided an impetus for two parallel and independent studies. The first used dissociated single neurons cultured from the neural tube of embryonic *Xenopus*, in which the effects of an applied electric field could be assessed on individual neurites and growth cones [9,10]. Since the electric field did not determine the point of origin of a neurite from its soma, neurites sprouted in random directions. Those growing roughly parallel to the field, grew preferentially towards the cathode, while those which had sprouted perpendicular to the field turned to grow cathodally (Fig. 1). Cathodally directed growth was a field-strength-dependent event (Fig. 2). The threshold level for cathodal-directed reorientation was very low, 7 mV/mm, corresponding to an external voltage gradient of less than 0.5 mV across a 50-μm-diam. growth cone [10]. It is interesting that reorientation is a heterogeneous affair. Gradual continuously curving neurites are often seen side by side with processes which have made sharp, right-angled changes in directions to grow cathodally (e.g. Fig. 1). Neurites also selectively branch cathodally in an applied electric field (see below). This can occur either by growth cone bifurcation, or by back-branching along the neurite shaft. Perhaps the combined effects of growth cone reorientation, growth cone bifurcation and neurite back-branching give rise to the array of different reorienting profiles observed. Alternatively, the heterogeneous response may reflect the heterogeneity of the culture population. These early *Xenopus* embryonic neural tube cultures contain predominantly cholinergic, motoneurons (up to 70–80%). Other cell types, however, include sensory Rohon–Beard neurons [11].

In this context also, it is worth noting a major difference between this study and that of Jaffe and Poo [8,10]. In the former, neurites showed differential cathodal-directed growth and turning; in the latter, no turning was seen, only faster growth cathodally. Again, one interesting explanation may be that different cell types respond differently to a small applied electric field: frog motoneurons turn cathodally, chick sensory neurons do not. Neurons show differential sensitivity to many extracellular guidance cues, depending on whether or not they possess the requisite receptors (e.g. neurotrophins, neurotransmitters, adhesion molecules etc). The same may apply to endogenous DC electric fields.

Fig. 1.   *Xenopus* **neural tube cells grow cathodally in a small applied electric field (170 mV/mm)**

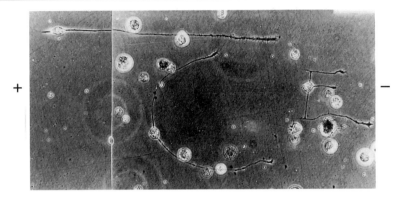

The field vector runs horizontally with polarity as shown (cathode at right). Three types of response are evident; sharp, right-angled turning towards the cathode; gradual curving towards the cathode; and differential growth, with cathode-facing neurites longer than anode-facing neurites. Cell bodies are approx. 30 μm in diameter. (Reproduced from Plate 1E [10]).

To prevent the build up of any chemical gradients or bulk flow of medium arising in response to current flow, culture medium was perfused continuously through the experimental chambers, at right angles to the electric field vector. Differential cathodal growth and cathodal turning were quantitatively identical in the presence or absence of cross-perfusion. Serial photography of individual cells established that turning responses occurred without movements of the cell body, while the substrate was a non-polarizable tissue-culture plastic, whose macromolecular structure could not be reorganized by such weak electric fields. These considerations, together with more recent information regarding mechanisms and interactions (see below), clearly indicate that nerves reorient as a direct response to an electric field alone.

Much of the work of Hinkle *et al.* [10] was confirmed and extended by a parallel study using dissociated cells from the neural tube of *Xenopus* [12]. Importantly, this study also provided the first clues to mechanisms, and implicated cell-surface receptors in the reorientation response (see below). Neurites also turned cathodally in a uniform pulsed field or if current was applied focally at the growth cone from a micropipette [13]. Presentation of the field in this form was thought to more closely resemble the fields which a nerve may encounter in the CNS or at peripheral synapses respectively.

There is additional phenomenology. Not only do neurites grow slower anodally, many actively retract and may reabsorb completely [10,14]. Anode-directed neurite retraction can be reversed by switching field polarity [14]. Although the importance of inhibitory influences on nerve growth has been appreciated for some time [15], there are no studies of the mechanisms underlying these growth-suppressive effects of electric fields. Nonetheless, they add to the repertoire of different ways in which a single cue may sculpt neuronal morphology. Neuronal form can be determined still further by directed branching. An electric field applied focally from a micropipette placed alongside the shaft of a *Xenopus* or *Helisoma* neurite causes filopodial-like extensions to appear on the cathodal side; some thicken, develop growth cones and become established branches [16,17]. Both Hinkle et al. [10] and Patel and Poo [12] report an increase in the proportion of neurons sprouting processes in field-exposed as opposed to control cultures. Thus, electric fields may stimulate neuronal differentiation in these cultures. Again, the mechanisms remain unexplored.

Fig. 2.	Composite line drawings of neurite growth in the absence (a) and presence (b–e) of small DC electric fields (50–200 mV/mm)

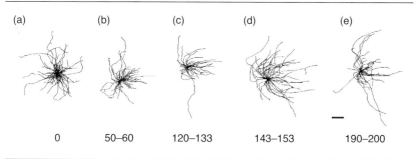

Enlarged tracings were made of the outgrowth from individual neurons and the composite pictures produced by superimposing their cell bodies. (**a**) Control neurite observed for 5 h in the absence of an electric field. Neurites have grown out from all parts of the cell body and show random, non-directed growth. (**b–e**) Neurite orientation after 5 h exposure to applied fields of: **b**, 50–60 mV/mm; **c**, 120–133 mV/mm; **d**, 143–153 mV/mm; **e**, 190–200 mV/mm. At all field strengths a number of neurites have turned to grow towards the cathode. However, in applied fields stronger than 143 mV/mm, significantly more neurites were oriented than at lower field strengths (compare **b** and **c** to **d** and **e**). The field vector runs horizontally, with the cathode at right. Scale bar = 100 μm.

In summary, the array of effects which a small DC field of physiological magnitude has on neurons includes: (1) directed growth; (2) directed branching; (3) directed retraction; (4) directed rates of growth; and (5) more neurons sprouting neurites. Taken together with the observations that different nerves may respond differently [4] and that electric fields interact with other cues (positively and negatively), in ways outlined below, it is clear that what at first sight seems to be a unidirectional cue, may in fact be much more elaborate and subtle in its influence on any given neuron.

Electric fields interact with co-existing guidance cues

A list of other types of guidance cues, some of which are dealt with in this volume includes neurotransmitters, adhesion molecules of the extracellular matrix (ECM) and cell surfaces, proteoglycans (PGs) (similarly distributed) and chemical gradients (e.g. of neurotrophins such as nerve growth factor). We have evidence that a small applied electric field interacts with examples of each of these guidance cues to modulate guided nerve growth.

Neurotransmitters modulate nerve growth and guidance and interact with electric field-directed nerve guidance

Embryonic growth cones spontaneously release neurotransmitter before making target contact [18,19]. The functional significance of this is not known. In the section Mechanisms of nerve reorientation in an electric field (below), a role for asymmetric redistribution of the neuronal nicotinic acetylcholine (ACh) receptor (AChR) is outlined. Here, suffice to say that *Xenopus* neurons possess both muscarinic and nicotinic AChRs [20], and that the *Xenopus* growth cone spontaneously releases ACh [18]. Neurotransmitters modulate nerve growth. 5-Hydroxytryptamine (serotonin) for example, selectively inhibits motility of specific *Helisoma* growth cones, while the nicotinic antagonist d-tubocurarine enhances process outgrowth from rat retinal ganglion neurons [21,22]. We tested the hypothesis that neurotransmitter receptors are involved in electric field-induced nerve reorientation, by combining field exposure with pharmacological inhibitors of neurotransmitter-activated receptors. *Xenopus* neurites, which spontaneously release ACh from their growth cones, do not turn in an electric field if the nicotinic AChR antagonist d-tubocurarine is also present (Fig. 3) [23,24]. In an electric field plus the muscarinic AChR antagonist atropine, *Xenopus* neurites turned faster and to a greater extent cathodally. A similar marked enhancement of turning occurred in the presence of the P2-purinergic receptor inhibitor, suramin (Fig. 4). By inference therefore, nicotinic AChR activation may be essential for cathodal reorientation, muscarinic AChR activation may prevent cathodal turning, while ATP (which is co-stored and co-released with ACh in many neurites) may interact with a third receptor type, a purinoceptor, to inhibit cathodal reorientation [24] (Fig. 5). There appears therefore to be a complex interplay between activated neurotransmitter receptors and a small applied electric field, which determines the extent of nerve guidance *in vitro*.

Fig. 3. (Top) Composite line drawings of field-treated neurite
outgrowth in the absence and presence of d-tubocurarine
(0.1 μM) and (bottom) cumulative angle turned by a
population of *Xenopus* neurites and the inhibitory effects of
0.1 μM d-tubocurarine

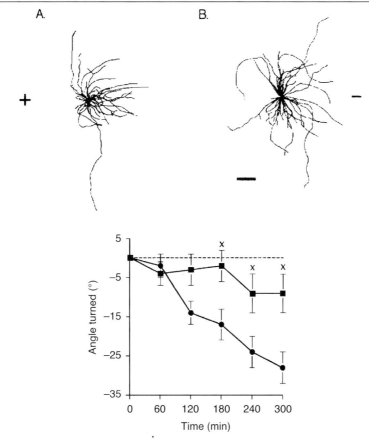

(Top) In an electric field alone (A), the majority of neurites have turned to grow cathodally, whereas after
addition of 0.1 μM d-tubocurarine (B) randomly directed growth occurred. The field vector runs from left
to right with the cathode at the right. Field strength = 50–133 mV/mm. Scale bar = 100 μm. (Bottom)
The extent of growth cone deflection was assessed hourly during exposure to an applied electric field
(50–133 mV/mm) in the absence (circles) and presence of the nicotinic AChR antagonist d-tubocurarine
(0.1 μM). Negative angles represent turning towards the cathode, positive angles, turning towards the
anode. After 5 h, the mean angle turned by neurites was −28 ± 4 ° (n = 103 neurites) in the electric field
alone. In neurites exposed to 0.1 μM d-tubocurarine and an applied field (squares), cathodal turning was
inhibited substantially, with neurites only turning through a mean angle of −9 ± 7 ° (n = 33 neurites) in
5 h. ˣP < 0.05.

Fig. 4.   **(a) Mean angle (± S.E.M.) turned by neurites in the absence (circles) and presence (squares) of atropine (1 or 50 μM) and (b) enhancement of neurite orientation by suramin**

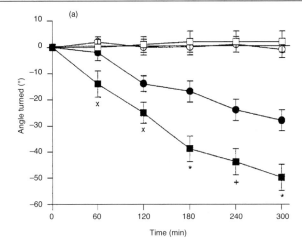

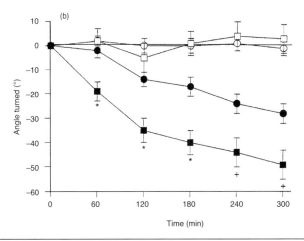

(a) In control cultures (no electric field; open symbols) neurites in normal or atropine-containing medium turned neither to the left or right of the culture chambers. In an applied electric field (50–133 mV/mm; filled symbols), neurites turned cathodally. Atropine (1 or 50 μM; filled squares) almost doubled the extent of cathodal reorientation seen in normal medium (filled circles). *P < 0.001; +P < 0.01; ×P < 0.05 compared with electric field alone. Atropine also induced faster cathodal turning of neurites. Negative angles represent turning towards the cathode (right in controls), positive angles, turning towards the anode (left in controls). Numbers of neurites = 69–179. (b) Open symbols, control (no electric field); filled symbols, plus applied electric field of 50–133 mV/mm. Circles, normal medium; squares, plus 38.5 μM suramin, a P2-purinoceptor antagonist. *P < 0.001; +P < 0.01 compared with field alone. Numbers of neurites = 57–179.

Fig. 5.   **Effects of atropine and suramin on cathodal reorientation of**
                    ***Xenopus* neurites**

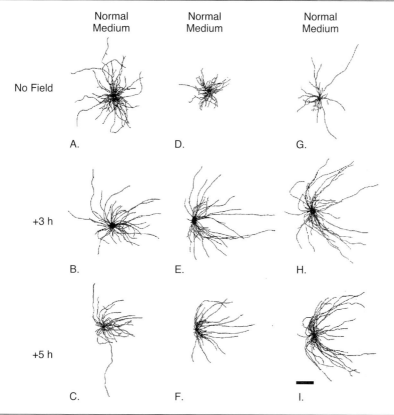

*Composite line drawings of neurites in normal medium (**A–C**), 50 μM atropine (**D–E**), or 38.5 μM*
*suramin (**G–I**). **A, D** and **G**, control neurites grown for 5 h in the absence of an electric field. Neurites*
*have grown out from all parts of the cell body and show random, non-directed growth. **B, E** and **H**, neurite*
orientation after 3 h exposure to applied fields of 50–133 mV/mm. In an applied field alone (B) little
orientation has occurred, neurites turning only −17 ± 4 ° (n = 103), whereas neurites in field plus atropine
*(**E**) or field plus suramin (**H**) show marked cathodal orientation, neurites having already turned −39 ± 5 °*
(n = 67) and −40 ± 5 ° (n = 57) respectively (P < 0.001 compared with field alone). After 5 h field
*exposure (**C, F, I**) stronger orientation is evident under all conditions. Significantly more neurites have*
*turned cathodally in the presence of atropine (**F**) and suramin (**I**) than in the field alone (**C**). Overall*
angles turned after 5 h were −28 ± 4 ° (n = 103), normal medium; −50 ± 5 ° (n = 69), atropine;
*−49 ± 6 ° (n = 57), suramin. Scale bars: 50 μm, **B, E** and **H**; 100 μm, **A, C, D, F, G** and **I**. The field*
vector runs horizontally with the cathode at right.

Intriguingly, a gradient of neurotransmitter alone can reorient a growth cone. A
focal source of ACh released in a pulsatile manner from a micropipette tip

establishes a gradient of transmitter. *Xenopus* growth cones sense this and reorient to grow up the gradient within a matter of minutes; a response which involves an elevated intracellular calcium concentration ($[Ca^{2+}]_i$) [25]. Here then is the first example of an electric field interacting with a 'respectable' guidance cue. Neurotransmitters and electric fields, which alone can guide nerve growth, interact *in vitro* to up- or down-regulate field-induced nerve guidance.

Two other aspects of electric field-induced nerve growth are influenced by AChR activation. *Xenopus* neurites grow faster cathodally than anodally. They also branch more frequently in an applied electric field than in control (no field) cultures, while 80% of these branches are directed cathodally [26]. Neither of these field-induced events occurs in the presence of the nicotinic AChR antagonist d-tubocurarine, which imposes equal rates of growth on cathodal- and anodal-directed neurites and an even distribution in the direction in which branches project [24]. Thus, in addition to the direction of nerve growth, both the rate at which nerves grow and the frequency and positioning of side branches is under the dual, interactive control of a neurotransmitter and an applied electric field. The finding that neuronal nicotinic AChR activation is a necessary step for field-directed nerve branching is novel and begs the question of whether other cues which direct nerve branching also may operate via neurotransmitter receptor activation. For example, back-branches from rat corticospinal projections are induced and grow into the pons, in response to a locally produced chemotropic agent [27]. Perhaps this chemotropic branching signal involves surface neurotransmitter receptor activation as an initiating step.

Localized neurotransmitter release and endogenous electric fields are likely to co-exist during development and therefore similar interactions may occur *in vivo*. Taken together, these studies indicate several possible functional roles for neurotransmitter release from a growth cone. Prior to reaching its target, the growth cone may release the contents of neurotransmitter-containing vesicles in order to modulate the growth rate, branching pattern and/or direction of growth of itself, or of neighbouring neurites. The interactions with an electric field indicate that a self-feedback system does occur, where ACh is released by the growth cone, in order to influence its own direction of growth (see Mechanisms of nerve reorientation in an electric field below).

Substrate interactions determine the direction of field-induced nerve guidance

Many elements of the ECM, or indeed molecules bound to cell surfaces, carry a net charge of variable magnitude, most frequently a negative charge. Variations in net surface charge alter both field-induced membrane protein mobility and the conductance properties of various voltage-gated ion channels (see Mechanisms of nerve reorientation in an electric field below). Since these two phenomena may be central to the mechanisms underlying field-induced reorientation, modulating them may influence the orientation of nerves in an electric field. Much of the preceding phenomenology has dealt with embryonic *Xenopus* neurites growing on a tissue-culture plastic substrate, in a serum-supplemented medium. However, the same neuronal type grown on tissue-culture plastic coated with the highly polycationic poly-L-lysine reorients anodally, that is in the opposite direction [28,29]. More recent unpublished work indicates that the greater the charge on the polylysine, the greater the extent of anodal reorientation (A.M. Rajnicek and C.D. McCaig, unpublished work). Thus surface-charge-related effects strongly influence the basic direction in which an electric field reorients a growth cone. How does this arise? Can one manipulate cell-surface charge in ways which predictably alter the direction of field-induced reorientation? And should we expect variations in surface charge *in vivo* to interact with endogenous electric fields?

There is good evidence using fluorescent probes that receptors for the plant lectin concanavalin A (Con A), or indeed AChRs, move within the plane of the plasma membrane and accumulate predominantly at the cathodal pole on field-exposed neurons and myoblasts (no polylysine present) [12,30]. This paradoxical accumulation of negatively charged receptors at the negative pole of a cell, occurs because of the electrophoretic displacement of positively charged counter ions, e.g. Na^+ and K^+, along with associated water molecules in the immediate vicinity of the membrane. It is this movement of water which sweeps receptors cathodally by electro-osmosis, despite their net negative charge [31]. Among the interventions which alter the balance between electro-osmosis and electrophoresis, and therefore the direction of receptors migration and accumulation, is the presence of polycations (e.g. polylysine), bivalent cations, or polyanions. Polycations decrease membrane surface charge by binding and screening negative elementary charges of the cell surface. This directly inhibits electro-osmosis and field-induced receptor migration [32]. Blocking the development of receptor asymmetry, by preincubating cells with polycationic Con A (which cross-links and immobilizes receptors bearing glucose and mannose residues), also prevents nerves from reorienting cathodally [12]. Decreasing membrane surface charge additionally increases the depolarization required to activate voltage-gated ion channels and inhibits channel conductance [33]. Thus it is possible to test the role of surface charge in field-induced growth cone reorientation by systematically adding to the culture medium polycations, which should inhibit or reverse cathodal-directed reorientation of growth cones [26], or polyanions, which should enhance cathodal turning [34]. Such experimental manipulations of cell-surface charge did modulate field-induced neuronal guidance, but not in a straightforward and predictable manner. While polycations, such as the aminoglycoside antibiotic neomycin or polylysine itself, either inhibited or reversed directional reorientation, as would be predicted from the effects of reducing membrane surface negative charge; lyotropic anions, which increase membrane surface charge, either had no effect, or inhibited cathodal reorientation. Excess perchlorate and thiocyanate had no effect, while excess sulphate inhibited cathodal turning [26,34]. In addition to its effects on surface charge, SO_4^{2-} increases production of the second messengers diacylglycerol (DAG) and inositol trisphosphate ($InsP_3$). Interestingly, excess lithium which also modulates polyphosphoinositide metabolism had almost identical effects to SO_4^{2-} on field-induced neurite turning [34]. Increasing surface charge with lyotropic anions therefore neither enhanced galvanotropic reorientation, nor did it underlie the inhibitory effects of sulphate (since perchlorate and thiocyanate were without effect). We suspect that the inhibitory effects of SO_4^{2-} may be due to changes in the inositol phospholipid second messenger system. This signalling pathway may be involved also in the inhibitory effects of the aminoglycoside antibiotics; neomycin, tobramycin and gentamicin all prevent cathodal reorientation and in addition to reducing cell-surface charge are known to inhibit phospholipase C (PLC) in several cell types [35].

Given the complex effects which altering cell-surface charge has on field-directed neurite growth, we speculate that the effects of known endogenous electric fields in directing nerve growth may vary between regions expressing differently charged molecules, regionally and temporally. These are issues which have not been addressed to date. One such molecule could be the negatively charged polysialic acid which in association with the adhesion molecules N-CAM and L1 appears to regulate axonal fasciculation and pathfinding of chick motoneurons as they undergo a sorting process in the plexus region [36].

Proteoglycans (PGs) modulate field-induced nerve guidance

Many other elements on cell surfaces and within the ECM carry net negative charges, for instance the sulphated glycosaminoglycans (GAGs) and PGs. In their own right, such molecules can direct nerve growth. They also interact with electric fields to modulate neurite guidance.

A diverse and growing set of PGs is expressed in the nervous system; many are regulated developmentally. A heparan sulphate PG (HSPG) is thought to be essential for normal pathfinding in the cockroach central nervous system (CNS); its inhibition disrupts the directed growth of pioneer axons [37]. Chondroitin sulphate PGs (CSPGs) have been implicated both as growth promoters and growth inhibitors. Outgrowth of embryonic rat neocortical and mesencephalic neurites is enhanced by specific CSPGs. Intriguingly, it is the core protein of the PG which stimulates neocortical cell growth, while the GAG side-chains alone promote mesencephalic neurite growth [38,39]. A CSPG is also expressed in axon-free regions of the developing retina in chick and rat. The area expressing PG moves peripherally with time, but is always located at the outer edge of the growing axons [40]. By inhibiting nerve growth, it is thought that this CSPG may play an essential role in pathfinding in the retina, a proposal supported by the observation that disrupting CSPG enzymically with chondroitin ABC lyase disrupts the normal pattern of retinal ganglion cell (RGC) growth towards the optic fissure [41].

Thus there is a growing list of both the developmental expression of PGs in the embryonic nervous system and of the positive and negative effects of PGs on nerve growth and guidance. Additionally, neurotransmitter receptor activation in fetal rat hippocampal neurons promotes synthesis and release of HSPGs with neurite growth-promoting properties [42]. This may be important in the context of the central role proposed for neurotransmitter receptor activation in field-induced nerve guidance (see Neurotransmitters modulate nerve growth and guidance and interact with electric field-directed nerve guidance above, and Mechanisms of nerve reorientation in an electric field below). What then are the interactions between PGs and an applied electric field of physiological magnitude? To date, we have examined two commercially available PGs, which share many of the properties of those found in the embryonic nervous system. These are a bovine nasal cartilage PG (BNC-PG) and a rat chondrosarcoma PG (RC-PG). Importantly, their GAG side chains differ markedly. About 70% of the side chains of BNC-PG are chondroitin sulphate (CS) the rest are keratan sulphate. Moreover, 80% of the CS GAG chains are sulphated in the 6 position, with only 20% sulphated in the 4 position on the GAG side chain. By contrast, RC-PG contains around 100 CS GAG chains, all are sulphated in the 4 position.

The growth of embryonic *Xenopus* neurites is directed randomly in the presence of 0.1 µg/ml of either PG alone. In an applied electric field of 50–133 mV/mm, 62% of neurites turned cathodally (64/103) and after a 5 h culture period, had turned through a mean angle of -28 ± 4 ° ($n = 103$). Exposure of neurites to either PG in the presence of an electric field substantially modulated field-induced guidance. When exposed to 0.1 µg/ml of BNC-PG plus an electric field, a greater proportion of neurites turned cathodally (82%; 27/33); they also turned to a greater extent (mean angle after 5 h of -49 ± 8 ° ($n = 33$) (L. Erskine and C.D. McCaig, unpublished work). In contrast, in cultures containing 0.1 µg/ml of RC-PG, galvanotropism was inhibited. Both the proportion of neurites turning cathodally (33%; 10/30), and the overall angle turned towards the cathode (-9 ± 6 °; $n = 30$) were reduced significantly compared with the effects of the applied field alone ($P < 0.01$). Thus, BNC-PG augmented, while RC-PG inhibited galvano-tropism.

Whether this modulation of cathodal turning was due to the core protein or to the GAG side chains was tested by exposing neurites simultaneously to various isolated GAG chains plus an applied electric field. Remarkably, chondroitin-6-sulphate (C-6-S) GAG markedly augmented cathodal turning, to the same extent as seen with BNC-PG. In 10 µg/ml C-6-S, 85% (33/39) of neurites turned cathodally, and by 5 h in an applied field, there was a twofold increase in the mean angle turned towards the cathode (-57 ± 5 °). Chondroitin-4-sulphate (C-4-S) GAG (10 µg/ml), by contrast, substantially suppressed cathodal turning, again to a similar extent to that seen with RC-PG. Thus the same modulatory effects of these two PGs were seen when their major GAG side chains were tested alone, indicating that the core proteins may not be required for these guidance-modulating events. Given that these GAG chains involve multiple, repeat disaccharide units of either C-6-S or C-4-S, the effects of the disaccharide subunits alone were tested. The extent of galvanotropism was unaffected quantitatively by the presence of 10 μg/ml of either C-6-S, or C-4-S, when presented as disaccharide subunits in the culture medium.

This is intriguing and potentially very important. The extent of field-induced nerve reorientation can be up- or down-regulated by one or other of two naturally occurring PGs. Here again is an instance where one of a family of 'respectable' guidance cues, a PG, interacts with another guidance mechanism, an applied electric field, with resultant modulation of nerve guidance. Thus in any *in vivo* situation where specific PGs co-exist with an endogenous electric field (see Chapter 9), modulation of field-induced guidance should be considered. Moreover, these modulatory effects appear to be properties of repeat GAG chains alone. A CS GAG chain sulphated in the 6 position augmented field-induced cathodal reorientation, while a CS GAG chain differing only by virtue of being sulphated in the 4 position inhibited cathodal turning. Surface-charge considerations alone can not account for modulated guidance, since the C-6-S and C-4-S GAGs are similarly charged, but have opposing effects. Rather it is the repeating pattern of the position of sulphation which is critical, since modulation of guidance was lost when this repeat pattern was no longer present, i.e. when the disaccharide subunits rather than repeat GAG chains were presented along with the applied field. This implies that it is the spacing between charged groups which determines their different effects. Cations with charged groups in specific configurations inhibit the binding of herpes simplex virus to its cellular receptor [44]. Perhaps the functioning or ligand interactions of specific membrane proteins in the growth cone may be altered by binding specific GAGs. Among the growth cone proteins potentially involved could be neurotransmitter receptors, especially the neuronal nicotinic AChR and voltage-gated calcium channels (see Mechanisms of nerve reorientation in an electric field below).

Alternatively, CSPGs/CS GAGs might alter field-induced neurite orientation by their ability to bind certain growth factors [45]. Fibroblast growth factor (FGF) for instance, binds to HSPG. Not only is this thought to protect FGF from degradation, but it appears to be an essential step in presenting an active form of basic FGF (bFGF) to its own receptor [46]. PG binding also is thought to provide a matrix- or cell-surface-bound reservoir of FGF. Specific charge sequences in the GAG side chains mediate the binding of growth factors (e.g. FGF) to PGs. Thus it is conceivable that the different sulphation patterns and hence charge positionings of C-6-S and C-4-S GAGs, allow differing degrees of growth factor binding. Two candidate growth factors which could be involved are the neurotrophins brain-derived neurotropic factor (BDNF) and neutrotropin 3

(NT-3). These are present in motoneurons, modulate nerve growth and, as outlined below, enhance field-induced neurite orientation.

Neurotrophins modulate field-induced nerve guidance

Recently, we discovered that the nature of the foetal bovine serum (1%, v/v) added to all our *Xenopus* culture media was critical in determining the quantitative extent of field-induced cathodal reorientation. In some sera galvanotropism was weak and substantially greater field strengths were required to achieve standard levels of robust cathodal reorientation. Some element within the serum either inhibited reorientation, or was an essential co-requisite although only present in small amounts in some batches. Growth factors were potential candidates.

The neurotrophin family of growth factors have widespread effects in promoting neuronal survival and differentiation. Additionally, a neurotropic capacity has been demonstrated *in vitro* for NGF which, in gradient form, reorients DRG neurites [47], although a physiological tropic role for NGF (or the other neurotrophins) has not been demonstrated developmentally [48]. One of the functions of NT-3 and BDNF is to potentiate both spontaneous and evoked release of neurotransmitter at developing neuromuscular synapses in *Xenopus* co-cultures [49]. Given that neuronal nicotinic AChR activation is essential for field-induced cathodal reorientation (see Neurotransmitters modulate nerve growth and guidance and interact with electric field-directed nerve guidance above) and that growth cone release of neurotransmitter may be central to the mechanism of field-induced reorientation (see below), we have tested the hypothesis that by increasing ACh release, NT-3 and BDNF may augment field-induced cathodal reorientation. For both neurotrophins this was found to be the case [50].

NT-3 increased both the rate and the overall extent of cathodal reorientation and reduced the threshold field strength required to produce a turning response. After 5 h in 100 mV/mm (without NT-3 present), neurites had only turned to a small extent, -9 ± 5 ° ($n = 79$). By contrast, in the continued presence of 50 ng/ml, neurites had turned in excess of -20 ° within 1 h at 100 mV/mm, and by 5 h had turned through a mean angle of -37 ± 6 ° ($n = 68$). In this new foetal bovine serum, the threshold level for responding with cathodal reorientation lies somewhere around 100 mV/mm, at which neurites only turn through a mean of -9 ± 5 ° ($n = 79$). After addition of NT-3 (100 ng/ml), however, neurites showed a more robust turning response (-16 ± 7 °; $n = 58$) to a field of half the magnitude (50 mV/mm) [50]. Although experiments are at a preliminary stage, it appears that BDNF may be even more effective than NT-3 in enhancing field-induced cathodal turning and in resetting the threshold value for a response to a lower field strength.

NT-3 and BDNF may be available to motoneuronal growth cones from target tissues, or may be supplied by release from an intrinsic source, or from neighbouring glia. Where any of these sources coincide with endogenous electric fields, interactions which enhance the efficacy of field-induced nerve guidance are to be expected.

Mechanisms of nerve reorientation in an electric field

Cathodal reorientation of *Xenopus* neurites can be modulated substantially therefore by at least four types of naturally occurring influence: neurotransmitter receptor activation, extracellular surface charge, PGs and their associated GAG chains, and growth factors of the neurotrophin family. Each of these local environmental cues on their own may have specific receptor interactions and downstream second messenger consequences. Might the same be true of an applied (or endogenous) electric field of physiological magnitude? Also what can be said of the mechanisms underlying these various interactions? Partial explanations are emerging.

DC electric fields impinge on cells in two ways [32,51]. First, because of the very high resistance of the plasma membrane the vast majority of a potential drop across a cell is experienced outside the cell, with almost no internal voltage gradient arising. Electric fields of the order of 100 mV/mm produce a voltage gradient across a growth cone spanning 25–50 μm of 2.5–5 mV. This is sufficient to move charged integral membrane proteins. As outlined above, receptors for the lectin Con A, AChRs and the epidermal growth factor receptor are among many which have been demonstrated directly, using fluorescent probes, to migrate and redistribute asymmetrically on the cathodal side of muscle nerve or fibroblast-like cells. (The reasons for cathodal accumulation of negatively charged receptors have been considered above). Moreover, for *Xenopus* neurites exposed to an electric field, it is possible to prevent cathodal redistribution of surface receptors, by preincubating neurons with Con A before field application. By cross-linking receptors, Con A greatly reduces their lateral mobility. Neurites prevented from establishing receptor asymmetry in this way do not reorient cathodally, although they do continue to grow [12,52]. Thus there is an assumption that a key element in the mechanism whereby an electric field promotes directed growth involves surface receptor movement and the establishment of receptor asymmetry. Importantly, this has not been tested directly. Con A has many other effects on cells, including inhibition of neuronal nicotinic AChR functioning. It remains possible that this or some other effect might be the causal effect in inhibiting cathodal reorientation and that induced cathodal asymmetry of receptors is simply an epi-phenomenon. Despite this cautious caveat, a working model which can account for cathodal reorientation has, as a central element, induced receptor asymmetry [24]. An additional key element in the model is that motile growth cones spontaneously release neurotransmitter before ever making target contacts [18,19]. The functional significance of this is not known. Are these growth cones signalling to neighbouring cells, or to themselves, or indeed to both? We propose that for *Xenopus* growth cones, which release ACh, a part of the function of this release may involve a feedback, self-signalling mechanism to redirect growth [24]. ACh released spontaneously from the growth cone will be capable of activating growth cone neuronal nicotinic AChRs. In an electric field (endogenous or applied), there will be many more neuronal nicotinic AChRs (and other receptor types) on the cathodal-facing side of the growth cone. Neuronal nicotinic AChR activation will thus be greater cathodally. Neuronal nicotinic AChRs are highly permeable to Ca^{2+} [53] and can also increase $[Ca^{2+}]_i$ by depolarization-induced recruitment of voltage-dependent calcium channels (VDCCs) [54]. Thus $[Ca^{2+}]_i$ is likely to be greater on the cathodal than the anodal side of a field-exposed growth cone. In further support of this, calcium influx associated with neuronal nicotinic AChR activation is essential for ACh-induced growth cone reorientation [25]. Also, field-induced cathodal reorientation is partially inhibited by blocking VDCCs, or by disrupting calcium release from intracellular stores [55]. Among the effective inhibitors of cathodal turning are ω-conotoxin and ω-agatoxin GVIA. Respectively, these inhibit N-type and P-type VDCCs (both of which are heavily implicated in the mechanism underlying neurotransmitter release), which we are proposing are essential for field-induced reorientation. In this scheme, it is this imposed asymmetry of $[Ca^{2+}]_i$ in the growth cone which results in reoriented growth through its roles in asymmetric microfilament polymerization and membrane addition.

Of course elevating calcium may have consequences for other downstream second messengers. Calcium entry through VDCCs can directly activate PLC, with resulting increases in $InsP_3$, DAG and protein kinase C. The DAG lipase pathway for instance has been referred to already in this volume in the context of promoting neuronal outgrowth (see Chapter 3) and the involvement of this in field-directed outgrowth is untested. We have pharmacological evidence indicating the involvement of several downstream elements. The aminoglycoside antibiotics neomycin, gentamicin and tobramycin inhibit PLC and all prevent cathodal reorientation [26]. The strikingly similar inhibitory effects of sulphate and lithium suggest that the generation of $InsP_3$ is crucial to reorientation [34]. This is supported by the observation that disrupting calcium release from $InsP_3$-sensitive intracellular stores with thapsigargin partially inhibits turning, while the additional involvement of calcium-induced calcium release stores is indicated by the inhibitory effects of ryanodine which markedly reduces cathodal turning [55]. Finally, the relatively non-specific protein kinase C inhibitors sphingosine and H-7 both inhibit cathodal reorientation [23]. Thus there is a complex array of receptor and second messenger events which may form multiple parallel mechanisms whereby the effects of an extracellular field are transduced into directed growth. Where these pathways are shared with other guidance cues, or where there is cross-talk between simultaneously activated pathways, functional interactions would be likely.

The second possibility involves alteration to the membrane potential in cells exposed to DC fields. In a field of 100 mV/mm (physiological magnitude), the cathodal-facing membrane of a cell of 20–30 μm diameter will depolarize by around

2 mV. At much higher field strengths, fluorophores sensitive to membrane potential have been used to show directly that cathodal-facing membranes are depolarized by several tens of millivolts [56]. In this and a separate study, the level of depolarization was sufficient to activate VDCCs, since depolarization was accompanied by cathodally elevated $[Ca^{2+}]_i$ in the growth cone [56,57]. Thus while high field strengths (above 1 V/cm, which is an order of magnitude greater than those which reorient *Xenopus* neurites) do cause an asymmetric rise in $[Ca^{2+}]_i$, which is spatially localized to the cathodal-facing membrane and which causes an associated localized filopodial outshoot [57], depolarizations of around 2 mV (resulting from physiological field strengths) will have little or no direct effect on increasing the conductance of VDCCs.

Since our favoured model involves activation of asymmetrically distributed neuronal nicotinic AChRs in the growth cone, with downstream elevated $[Ca^{2+}]_i$, substances which alter either neurotransmitter release, or presynaptic calcium entry, might be expected to interfere with field-induced reorientation. One such substance is cannabis, or more accurately its major psychotropic component, Δ^9-tetrahydrocannabinol (Δ^9-THC). At a concentration of 100 nM Δ^9-THC completely inhibited cathodal reorientation of *Xenopus* neurites [58]. This may be a receptor-mediated event. The enantiomeric pair of cannabinoid analogues HU210 and HU211 show stereo-selective responsiveness. The active enantiomer HU210 also totally blocked cathodal turning at 1 nM, while the inactive HU211 was without effect even at a 100-fold higher concentration [58]. Not only might these results relate to the mechanism of field-induced nerve guidance, they may have a broader significance. There is recent evidence that early disruption of L1-mediated axon guidance may result in mental retardation [59]. Three-year-old children exposed to high levels of cannabinoids at foetal stages also show learning and behavioural deficits [60]. Perhaps early disrupted axon guidance is involved.

Mammalian neurites respond to an applied electric field

Such extrapolation from *Xenopus* growth cones to the human CNS is a sizeable leap! Nevertheless, mammalian neuronal growth cones also respond to small applied electric fields *in vitro*. Three recent studies indicate a rich variety of responsiveness. After 12–24 h exposure to a small applied field, the longest processes of a population of embryonic rat hippocampal neurons (axons?) have developed a striking perpendicular response to the field vector [61]. Presumptive axons and dendrites of the same cell type respond differently to a focally applied field from a micropipette tip. Dendrites turned cathodally, while axons showed no response, even at much higher field strengths [4]. In addition to the field being focally applied rather than applied globally across the whole culture chamber, this latter study involved relatively brief field-exposures of up to 20 min, rather than exposure for many hours. Whether this can account for the differing observations between the studies, or whether it indicates a potentially important physiological parameter (period of field exposure) is unclear. Pheochromocytoma cells (PC12 cells) by contrast, had grown predominantly anodally after 48 h exposure to small DC fields [62]. The mechanisms underlying any of these responses remain unexplored. However, the diversity of response encourages the view that endogenous electric fields may provoke responses from mammalian neurons which are cell specific, process specific and which may vary according to the duration of exposure.

Concluding remarks

A reintegration of endogenous electric fields as potential players in shaping the developing nervous system is taking place. Endogenous fields exist in mammalian, chick and amphibian embryos. Their magnitude, and spatial and temporal distribution are appropriate for them to act as cues in the developing nervous system, while disrupting endogenous fields specifically disrupts CNS development, at least on a gross level (see Chapter 9). DC fields alone profoundly influence nerve growth and guidance and do so by impinging on specific receptors and by activating well-recognized second messenger systems. Additionally, they interact with other co-existent guidance cues in ways which markedly modulate nerve guidance *in vitro*. Similar interactions may be expected to occur *in vivo* and may have profound developmental effects.

We are grateful for the financial support of The Wellcome Trust and for a Henry Dryerre Scholarship (L.E.) from the Carnegie Trust for The Universities of Scotland.

References

1. Child, C.M. (1921) The Origin and Development of the Nervous System, University of Chicago Press, Chicago
2. Ariens Kappers, C.U. (1917) J. Comp. Neurol. 27, 261–298
3. Ingvar, S. (1920) Proc. Soc. Exp. Biol. Med. 17, 198–199
4. Davenport, R.W. and McCaig, C.D. (1993) J. Neurobiol. 24, 89–100
5. Weiss, P. (1934) J. Exp. Zool. 68, 393–448
6. Marsh, G. and Beams, H.W. (1946) J. Cell. Comp. Physiol. 27, 139–157
7. Davies, A.M. (1989) Nature (London) 337, 553–555
8. Jaffe, L.F. and Poo, M-M. (1979) J. Exp. Zool. 209, 115–128
9. Robinson, K.R. and McCaig, C.D. (1980) Ann. N.Y. Acad. Sci. 339, 132–138
10. Hinkle, L., McCaig, C.D. and Robinson, K.R. (1981) J. Physiol. 314, 121–135
11. Tabti, N. and Poo, M-M. (1991) in Culturing Nerve Cells (Banker, G. and Goslin, K., eds.), pp. 137–154, MIT Press
12. Patel, N.B. and Poo, M.-M. (1982) J. Neurosci. 2, 483–496
13. Patel, N.B. and Poo, M.-M. (1984) J. Neurosci. 4, 2939–2947
14. McCaig, C.D. (1987) Development 100, 31–41
15. Patterson, P.H. (1988) Neuron 1, 263–267
16. McCaig, C.D. (1990) J. Cell Sci. 95, 605–615

17. Williams, C.V., Davenport, R.W., Dou, P. and Kater, S.B. (1995) J. Neurobiol. **27**, 127–140
18. Young, S.H. and Poo, M.-M. (1983) Nature (London) **305**, 634–637
19. Hume, R.I., Role, L.W. and Fischbach, G.D. (1983) Nature (London) **305**, 632–634
20. Perrins, R. and Roberts, A. (1994) J. Physiol. **478**, 221–228
21. Haydon, P.G., McCobb, D.P. and Kater, S.B. (1984) Science **226**, 561–564
22. Lipton, S.A., Frosch, M.P., Phillips, M.D., Tauck, D.L. and Aizenman, E. (1988) Science **239**, 1293–1296
23. Ahmed, T., Erskine, L., Shewan, D.A., Stewart, R. and McCaig, C.D. (1992) J. Physiol. **446**, 42P
24. Erskine, L. and McCaig, C.D. (1995) Dev. Biol. **171**, 330–339
25. Zheng, J.Q., Felder, M., Connor, J.A. and Poo, M.-M. (1994) Nature **368**, 140–144
26. Erskine, L., Stewart, R. and McCaig, C.D. (1995) J. Neurobiol. **26**, 523–536
27. Heffner, C.D., Lumsden, A.G.S. and O'Leary, D.D.M. (1990) Science **247**, 217–220
28. Rajnicek, A.M., Cork, R.J. and Robinson, K.R. (1989) Soc. Neurosci. Abstr. **15**, 1036
29. Rajnicek, A.M., Robinson, K.R. and McCaig, C.D. (1994) J. Physiol. **481**, 25P
30. Stollberg, J.S. and Fraser, S.E. (1988) J. Cell Biol. **107**, 1397–1408
31. McLaughlin, S. and Poo, M.-M. (1981) Biophys. J. **34**, 85–93
32. Poo, M.-M. (1981) Annu. Rev. Biophys. Bioeng. **10**, 245–276
33. Green, W.N. and Andersen, O.S. (1991) Annu. Rev. Physiol. **53**, 341–359
34. Erskine, L. and McCaig, C.D. (1995) J. Physiol. **486**, 229–236
35. Schwertz, D.W., Kreisberg, J.L. and Venkatachalam, M.A. (1984) J. Pharm. Exp. Ther. **231**, 48–55
36. Tang, J., Rutishauser, U. and Landmesser, L. (1994) Neuron **13**, 405–414
37. Wang, L. and Denburg, J.L. (1992) Neuron **8**, 701–714
38. Iijima, N., Oohira, A., Mori, T., Kitabatake, K. and Kohsaka, S. (1991) J. Neurochem. **56**, 706–708
39. Lafont, F., Rouget, M., Triller, A., Prochiantz, A. and Rousselet, A. (1992) Development **114**, 17–29
40. Snow, D.M., Watanabe, M., Letourneau, P.C. and Silver, J. (1991) Development **113**, 1473–1485
41. Brittis, P.A., Canning, D.R. and Silver, J. (1992) Science **255**, 733–736
42. Sugiura, M. and Dow, K.E. (1994) Dev. Biol. **164**, 102–110
43. Reference deleted.
44. Langeland, N., Moore, L.J., Holmsen, H. and Haarr, L. (1988) J. Gen. Virol. **69**, 1137–1145
45. Ruoslahti, E. and Yamaguchi, Y. (1991) Cell **64**, 867–869
46. Yayon, A., Klagsbrun, M., Esko, J.D., Leder, P. and Ornintz, D.M. (1991) Cell **64**, 841–848
47. Gundersen, R.W. and Barrett, J.N. (1980) J. Cell Biol. **87**, 546–554
48. Davies, A.M., Bandtlow, C., Heumann, R. et al. (1987) Nature (London) **326**, 353–358
49. Lohof, A.M., Quillan, M., Dan, Y. and Poo, M.-M. (1992) J. Neurosci. **12**, 1253–1261
50. Zietlow, R., Stewart, R., Sangster, L. and McCaig, C.D. (1995) Soc. Neurosci. Abstr. **21**, 322.3
51. Jaffe, L.F. (1977) Nature (London) **265**, 600–602
52. McCaig, C.D. (1989) J. Cell Sci. **93**, 723–730
53. Verino, S., Amador, M., Luetje, C.W., Patrick, J. and Dani, J.A. (1992) Neuron **8**, 127–134
54. Pugh, P.C. and Berg, D.W. (1994) J. Neurosci. **14**, 889–896
55. Stewart, R., Erskine, L. and McCaig, C.D. (1995) Dev. Biol. **171**, 340–351
56. Bedlack, R.S., Wei, M.-D. and Loew, L.M. (1992) Neuron **9**, 393–403
57. Davenport, R.W. and Kater, S.B. (1992) Neuron **9**, 405–416
58. Tait, F., Kosterlitz, H.W., Pertwee, R.G. and McCaig, C.D. (1995) Soc. Neurosci. Abstr. **21**, 322.2
59. Wong, E.V., Kenwrick, S., Willems, P. and Lemmon, V. (1995) Trends Neurosci. **18**, 168–172
60. Day, N.L., Richardson, D.L., Goldschmidt, L. et al. (1994) Neurotoxicol. Teratol. **16**, 169–175
61. Rajnicek, A.M., Gow, N.A.R. and McCaig, C.D. (1992) Exp. Physiol. **77**, 229–232
62. Cork, R.J., McGinnis, M.E., Tsai, J. and Robinson, K.R. (1994) J. Neurobiol. **25**, 1509–1516

Index